T0269071

SPECTRAL THEORY AND DIFFERENTIAL OPERATORS

Already published

1 W.M.L. Holcombe *Algebraic automata theory*
2 K. Petersen *Ergodic theory*
3 P.T. Johnstone *Stone spaces*
4 W.H. Schikhof *Ultrametric calculus*
5 J.-P. Kahane *Some random series of functions, 2nd edition*
6 H. Cohn *Introduction to the construction of class fields*
7 J. Lambek & P.J. Scott *Introduction to higher-order categorical logic*
8 H. Matsumura *Commutative ring theory*
9 C.B. Thomas *Characteristic classes and the cohomology of finite groups*
10 M. Aschbacher *Finite group theory*
11 J.L. Alperin *Local representation theory*
12 P. Koosis *The logarithmic integral I*
13 A. Pietsch *Eigenvalues and s-numbers*
14 S.J. Patterson *An introduction to the theory of the Riemann zeta-function*
15 H.J. Baues *Algebraic homotopy*
16 V.S. Varadarajan *Introduction to harmonic analysis on semisimple Lie groups*
17 W. Dicks & M. Dunwoody *Groups acting on graphs*
18 L.J. Corwin & F.P. Greenleaf *Representations of nilpotent Lie groups and their applications*
19 R. Fritsch & R. Piccinini *Cellular structures in topology*
20 H Klingen *Introductory lectures on Siegel modular forms*
22 M.J. Collins *Representations and characters of finite groups*
24 H. Kunita *Stochastic flows and stochastic differential equations*
25 P. Wojtaszczyk *Banach spaces for analysts*
26 J.E. Gilbert & M.A.M. Murray *Clifford algebras and Dirac operators in harmonic analysis*
27 A. Frohlich & M.J. Taylor *Algebraic number theory*
28 K. Goebel & W.A. Kirk *Topics in metric fixed point theory*
29 J.F. Humphreys *Reflection groups and Coxeter groups*
30 D.J. Benson *Representations and cohomology I*
31 D.J. Benson *Representations and cohomology II*
32 C. Allday & V. Puppe *Cohomological methods in transformation groups*
33 C. Soulé et al *Lectures on Arakelov geometry*
34 A. Ambrosetti & G. Prodi *A primer of nonlinear analysis*
35 J. Palis & F. Takens *Hyperbolicity, stability and chaos at homoclinic bifurcations*
37 Y. Meyer *Wavelets and operators*
38 C. Weibel *An introduction to homological algebra*
39 W. Bruns & J. Herzog *Cohen–Macaulay rings*
40 V. Snaith *Explicit brauer induction*
43 J. Diestel, H. Jarchow & A. Tonge *Absolutely summing operators*

SPECTRAL THEORY AND DIFFERENTIAL OPERATORS

E. B. DAVIES

Department of Mathematics, King's College, Strand, London

PUBLISHED BY THE PRESS SYNDICATE OF THE UNIVERSITY OF CAMBRIDGE
The Pitt Building, Trumpington Street, Cambridge CB2 1RP, United Kingdom

CAMBRIDGE UNIVERSITY PRESS
The Edinburgh Building, Cambridge CB2 2RU, United Kingdom
40 West 20th Street, New York, NY 10011-4211, USA
10 Stamford Road, Oakleigh, Melbourne 3166, Australia

First published 1995
First paperback edition 1996

A catalogue record for this book is available from the British Library

Library of Congress cataloguing in publication data

Davies, E. B. (Edward Brian)
Spectral theory and differential operators / E. B. Davies.
p. cm.
Includes index.
ISBN 0 521 47250 4
1. Spectral theory (Mathematics). 2. Elliptic operators.
I. Title.
QA320.D317 1995
515′.7242–dc20 94-21008 CIP

ISBN 0 521 47250 4 hardback
ISBN 0 521 58710 7 paperback

Transferred to digital printing 2003

TAG

Contents

Preface

The theory of differential equations is one of the outstanding creations of the human mind. Its influence upon the development of physical science would be hard to exaggerate. The long history and many applications of the theory, however, make it almost impossible to write a balanced account of the subject. Thus authors of student texts are confronted with the choice between writing rather superficially on a range of topics or in more depth on some narrow field, in which they have a particular interest.

In this book I have given a simple introduction to the spectral theory of linear differential operators. This spectral theory is an outgrowth of fundamental work of David Hilbert between 1900 and 1910 on the analysis of integral operators on infinite-dimensional spaces – now called Hilbert spaces. However, like almost every important new development in mathematics, it was preceded by much related work, for example Poincaré's analysis of the Dirichlet problem and associated eigenvalues (1890–6). One could maintain that the subject started with the seminal work of Fourier on the solution of the heat equation using series expansions in sines and cosines, which was published by the Académie Française in 1822. Fourier submitted this work in 1807, during the Napoleonic era, and an account of his misfortunes during the fifteen year period before publication is given by Körner (1988). I have included the names and dates associated with a few of the key ideas in the text; a much more comprehensive account may be found in Dieudonné (1981).

Much of the subject matter in the book is confined not only to linear differential operators but even to second order elliptic differential operators. One justification for concentrating on this topic is that many of the equations which have proved important in the physical sciences and engineering over the last century involve operators of this type.

Most important among these is non-relativistic quantum theory, which is based upon the spectral analysis of Schrödinger operators. Applications of second order elliptic operators to geometry and stochastic analysis are also now of great importance. The constraints of time have forced me to omit any account of the wave equation, which certainly has an importance equal to what is included; it seems to me that this theory is best presented within the context of pseudo-differential operator theory, another vast subject. The theory of higher order elliptic operators extends the second order theory in an obvious sense, but loses the close connection which the second order theory has with Brownian motion. Without going into details, I can assure the student that the techniques presented in this book are of central importance to the higher order theory. Non-linear differential equations present a new order of variety and complication, but there also the linear theory is of importance. The KdV equation has solutions expressed in terms of properties of an associated family of linear problems. In many other non-linear problems the proof of local existence theorems depends upon the use of linear theory for operators with very weak assumptions on the coefficients.

Spectral theory is an extremely rich field which has been studied by many qualitative and quantitative techniques – for example Sturm–Liouville theory, separation of variables, Fourier and Laplace transforms, perturbation theory, eigenfunction expansions, variational methods, microlocal analysis, stochastic analysis and numerical methods including finite elements. Whether or not a student is going on to study one of these developments in detail, he or she should have an opportunity to see something of the underlying subject.

The book differs from other introductory texts on elliptic differential operators in that it does not assume that the operators have smooth coefficients, and does not depend upon a heavy use of embedding properties of Sobolev spaces. There is also more concern than in other texts at this level to establish relations between upper and lower bounds on the spectrum and quantitative assumptions about the regions in which the operators act. Many of the theorems are stated and proved in less than maximal generality, in order to make the essential ideas more accessible to students.

The prerequisites for reading the book are a knowledge of some Hilbert and Banach space theory, a course on Lebesgue integration and measure theory, and a little familiarity with Fourier transforms. I assume that the reader is aware of the elements of spectral theory for a bounded linear operator on an abstract Banach space. Chapter 2 presents an entirely new proof of the spectral theorem for unbounded self-adjoint operators. Although this proof is due to the author (Davies, 1993), it originates from a formula of Helffer and Sjöstrand which has had many important applications to spectral and scattering theory over the last few years. The proof is particularly straightforward and direct, and has the further advantage of using techniques which are of value elsewhere in the theory of differential equations. Those who already know a proof of the spectral theorem are not, however, disadvantaged.

The core chapters of the book are numbers 1, 2 and 4, all of which deal with abstract spectral theory; the remaining chapters are to some extent independent of each other, except that Chapter 8, on Schrödinger operators, depends heavily upon the treatment of constant coefficient operators using methods of Fourier analysis in Chapter 3. Most of the subject matter can be covered within a one year course by mathematics students in their fourth or fifth year at University. It is also possible to base a shorter course on the book, by taking Theorems 2.5.1, 4.5.2 and 6.2.3 as unproved statements of fact, and then following up their consequences. While almost all of the material has been known for a considerable time, I have presented it in a form which is related to my own research interests, concentrating particularly on Hilbert space techniques and the variational method. This method has been used for many decades, and has acquired extra strength in the last fifteen years with the increased development of the theory of quadratic forms.

It is possible to regard the present book as the first volume of a three-volume series by the author, the others being *One-Parameter Semigroups* (1980) and *Heat Kernels and Spectral Theory* (1989).

I would like to conclude by thanking the many people who have stimulated my interest in spectral theory over the last twenty years. I would also like to thank my students A. Arnal, O. Nicholas, M. Owen and P. Oleche for their constructive criticisms of the manuscript, and also for their help in eliminating a large number of minor errors. Needless to say I take full responsibility for those which remain.

Kings College, London E. Brian Davies
March 1994

1

The fundamental ideas

1.1 Unbounded linear operators

One of the key notions in any introductory course on functional analysis is that of a bounded linear operator. If A is such an operator on the Banach space $\mathcal{B}$ then there is a closed bounded subset $\mathrm{Spec}(A)$ of the complex plane called its spectrum. The proof that the spectrum is always non-empty is rather indirect, and this is related to the fact that the explicit determination of the spectrum of particular operators is often very difficult.

In this chapter we describe the appropriate context in which one can define and analyse the spectrum of unbounded linear operators, particularly those which are closed or self-adjoint. The description of the spectrum of particular operators will be the main focus of attention throughout the book.

Before one can start to study a differential operator one has to choose the Banach or Hilbert space in which it acts; we mention here that all Banach and Hilbert spaces in the book are assumed to be complex. It turns out that the spectrum of an operator can change depending upon the Banach space in which it acts. There is, however, another problem, namely that differential operators are unbounded when considered as acting on any of the usual Banach or Hilbert spaces. Because of this we cannot even start to study them until we have given a more general definition of a linear operator.

The key to this new definition is to drop the requirement that the domain of the operator is the whole of the Banach space in which the operator acts, and allow it to be a dense linear subspace. The precise specification of that subspace is very important since it turns out that the choice of different subspaces corresponds to the application of different boundary conditions to the same formal operator, which frequently leads to totally different spectra. For these reasons when using the term 'differential operator' we shall understand that we have already chosen the boundary conditions if we are thinking in more applied terms, or that we have already chosen the precise domain of definition of the operator if we are thinking in abstract terms.

Definition We define a linear operator on a Banach space $\mathcal{B}$ to be a pair consisting of a dense linear subspace L of $\mathcal{B}$ together with a linear map $A : L \to \mathcal{B}$. We call L the domain of the operator A and write $\mathrm{Dom}(A) := L$. If $\tilde{L}$ is a linear subspace of $\mathcal{B}$ which contains L and $\tilde{A}f = Af$ for all $f \in L$ then we say that $\tilde{A}$ is an extension of A.

A complex number λ is said to be an eigenvalue of such an operator A if there exists a non-zero $f \in \mathrm{Dom}(A)$ such that $Af = \lambda f$. Since the Banach spaces we are interested in are all spaces of functions, we call f an eigenfunction of the operator A. As in the more elementary theory of bounded linear operators, the set of eigenvalues is not to be confused with the spectrum (defined below), which is often a much larger set.

As an elementary example we choose $\mathcal{B}$ to be the space of all continuous functions on the interval $[a,b]$ and put $Af = -f''$ where $\mathrm{Dom}(A)$ is the set of all smooth (i.e. infinitely differentiable) functions on $[a,b]$. Every complex number is an eigenvalue of this operator, whose spectrum is therefore equal to $\mathbf{C}$. If, however, we take $\mathcal{B}$ to be the space of all continuous periodic functions on the interval $[a,b]$ and the domain to be the set of all periodic smooth functions on $[a,b]$, then the same formula defines a different operator with countable spectrum. The following example is more typical of those which we shall study later.

Example 1.1.1 We consider the operator H given formally by

$$Hf := -f'' \tag{1.1.1}$$

on the following alternative domains in the Hilbert space $L^2(a,b)$. To treat Dirichlet boundary conditions we take the domain L_D consisting of all twice continuously differentiable functions f on $[a,b]$ for which $f(a) = f(b) = 0$. To treat Neumann boundary conditions, however, we take the domain L_N of all twice continuously differentiable functions f on $[a,b]$ for which $f'(a) = f'(b) = 0$. Because we have two different domains the equation (1.1.1) determines two different operators which we shall call H_D and H_N. It is straightforward to determine the eigenvalues of these two operators and to see that 0 is an eigenvalue of H_N but not of H_D. In higher dimensions the difference between the spectrum of the Laplacian under Dirichlet and Neumann boundary conditions is much greater than in this example. □

In this chapter we shall illustrate our ideas by means of very simple operators such as H_D and H_N. The only reason for this is that we do not wish the reader to have to cope with the abstract theory and its applications at the same time. We ask the reader at this point to accept our reassurance that the conditions of the abstract theorems which we shall prove are actually verifiable in a wide range of more interesting applications.

The continuity of bounded linear operators is so useful that we need to have a replacement for it in our more general situation. This is provided by the notion of closedness. We will henceforth use the expression $\lim_{n\to\infty} f_n = f$ without further comment to mean that $\|f_n - f\|$ converges to zero as $n \to \infty$.

Definition Let A be an operator on $\mathcal{B}$ with domain L. We say that A is closed if whenever f_n is a sequence in L with limit $f \in \mathcal{B}$ and there exists $g \in \mathcal{B}$ such that $\lim_{n\to\infty} Af_n = g$, it follows that $f \in L$ and that $Af = g$.

There is an alternative formulation of this idea. The product $\mathcal{B}_1 \times \mathcal{B}_2$ of two Banach spaces $\mathcal{B}_1$ and $\mathcal{B}_2$ becomes a Banach space if we provide it with the norm

$$\|(f,g)\| := \{\|f\|^2 + \|g\|^2\}^{1/2}.$$

Other equivalent choices of the norm can be made, but one advantage of the above definition is that if $\mathcal{B}_i$ are both Hilbert spaces then $\mathcal{B}_1 \times \mathcal{B}_2$ is also a Hilbert space for the above norm. If we define the graph of A to be the set

$$\{(f,g) : f \in L,\, g \in \mathcal{B} \quad \text{and} \quad Af = g\},$$

then the operator is closed if and only if its graph is a closed subspace of $\mathcal{B} \times \mathcal{B}$.

The closed graph theorem states that if a closed operator has domain equal to $\mathcal{B}$, then it has finite norm. While conceptually extremely valuable, this result has the weakness of not giving any information about the size of the norm. We shall see below that the size of the norm of resolvent operators is important in locating the spectrum of A.

Definition If A is a linear operator on $\mathcal{B}$ with domain L then its spectrum $\mathrm{Spec}(A)$ is defined as follows. We say that a complex number z does not lie in $\mathrm{Spec}(A)$ if the operator $(z-A)$ maps L one-one onto $\mathcal{B}$, and the inverse (or resolvent) operator, which we shall denote by $R(z,A)$ or $(z-A)^{-1}$, is bounded.

The following lemma explains why the notion of closedness is so important. The statement and proof involve using analytic function theory for operator-valued functions of a complex variable. The definitions and proofs of the relevant results are simple adaptations of the corresponding results for complex-valued functions, and we leave readers to write out the details for themselves.

Lemma 1.1.2 *If the operator A does not have spectrum equal to the whole of the complex plane* $\mathbf{C}$ *then A must be closed. The spectrum* $\mathrm{Spec}(A)$ *of a linear operator A is always closed. More specifically let $z \notin \mathrm{Spec}(A)$ and let $c = \|R(z,A)\|$. Then the spectrum does not intersect the ball*

$$\{w \in \mathbf{C} : |z-w| < c^{-1}\}.$$

The resolvent operator is a norm analytic function of z and satisfies the resolvent equations

$$R(z,A) - R(w,A) = -(z-w)R(z,A)R(w,A), \tag{1.1.2}$$

$$R(z,A)R(w,A) = R(w,A)R(z,A), \tag{1.1.3}$$

$$\frac{\mathrm{d}}{\mathrm{d}z}R(z,A) = -R(z,A)^2, \tag{1.1.4}$$

for all $z, w \notin \mathrm{Spec}(A)$.

Proof Suppose that $z \notin \mathrm{Spec}(A)$ and let $B = (z-A)^{-1}$ be the inverse operator, which is bounded by hypothesis. Let $f_n \in \mathrm{Dom}(A)$, $\lim_{n\to\infty} f_n = f$, $\lim_{n\to\infty} Af_n = g$ and $h_n := (z-A)f_n$. Then

$$\lim_{n\to\infty} h_n = \lim_{n\to\infty}\{zf_n - Af_n\} = zf - g,$$

so

$$B(zf-g) = \lim_{n\to\infty}\{Bh_n\} = \lim_{n\to\infty}\{f_n\} = f.$$

This implies that $f \in \mathrm{Dom}(A)$ and $(z-A)f = zf - g$, or $Af = g$. Hence A is closed.

The remainder of the proof is very similar to the case when A is bounded. Consider the bounded operator C defined by

$$C := \sum_{n=0}^{\infty} (-u)^n B^{n+1}, \tag{1.1.5}$$

where the series is norm convergent if $|u| < \|B\|^{-1}$. The operator C satisfies the identities

$$C = B - uBC \quad , \quad C = B - uCB.$$

The first identity implies that the kernel $\mathrm{Ker}(C)$ and range $\mathrm{Ran}(C)$ of C satisfy

$$\mathrm{Ker}(C) \subseteq \mathrm{Ker}(B) \quad , \quad \mathrm{Ran}(C) \subseteq \mathrm{Ran}(B),$$

while the second implies

$$\mathrm{Ker}(B) \subseteq \mathrm{Ker}(C) \quad , \quad \mathrm{Ran}(B) \subseteq \mathrm{Ran}(C).$$

Since B has kernel $\{0\}$ and range $\mathrm{Dom}(A)$ we deduce that C is a bounded linear operator mapping $\mathscr{B}$ one-one onto $\mathrm{Dom}(A)$. If $f \in \mathrm{Dom}(A)$ and $g = (z - A)f$ then $f = Bg$, so $Cg = f - uCf$. Hence $C(z + u - A)f = f$. Since $(z+u-A)Cf = f$ for all $f \in \mathscr{B}$ by a similar calculation, we conclude that $C = (z + u - A)^{-1}$. This establishes both that $z + u \notin \mathrm{Spec}(A)$ if $|u| < \|(z - A)^{-1}\|^{-1}$ and that

$$(z + u - A)^{-1} = \sum_{n=0}^{\infty} (-u)^n (z - A)^{-(n+1)}. \tag{1.1.6}$$

The norm convergence of this series is more than enough justification for saying that the resolvent is a norm analytic function of z. The resolvent equation (1.1.4) is proved by differentiating the series (1.1.5) term by term, the justification for this being the same as for complex-valued power series. The resolvent equation (1.1.2) is proved by multiplying both sides by $(z - A)$, in which case it becomes an elementary identity. Upon interchanging w and z we see that (1.1.2) implies (1.1.3). □

It is possible to construct closed operators whose spectrum is either empty or equal to the whole complex plane. However, we shall mainly be interested in studying self-adjoint operators, whose spectrum is always a non-empty subset of the real line.

Although all of the above suggest that we should only study closed operators, there is a practical problem with this, namely that differential operators are usually defined initially on simple domains where they are not closed. This problem is overcome by yet more definitions.

Lemma 1.1.3 *An operator A on $\mathcal{B}$ with domain L is said to be closable if it has a closed extension $\tilde{A}$. In this case there is a closed extension $\overline{A}$, which we call its closure, whose domain is smallest among all closed extensions.*

Proof We define $\mathcal{D}$ to be the set of $f \in \mathcal{B}$ for which there exist $f_n \in \mathrm{Dom}(A)$ and $g \in \mathcal{B}$ such that $\lim_{n\to\infty} f_n = f$ and $\lim_{n\to\infty} Af_n = g$. Since $\tilde{A}$ is a closed extension of A it follows that $f \in \mathrm{Dom}(\tilde{A})$ and $\tilde{A}f = g$. Hence g is uniquely determined by f in the above situation. We define $\overline{A}f = g$ with $\mathrm{Dom}(\overline{A}) = \mathcal{D}$. Clearly $\overline{A}$ is an extension of A and every closed extension of A is also an extension of $\overline{A}$. The graph of $\overline{A}$ is the closure of the graph of A in the Banach space $\mathcal{B} \times \mathcal{B}$. Hence $\overline{A}$ is a closed operator. □

Many of the differential operators which we will study have the key property of self-adjointness. An intermediate but much more elementary property is that of being symmetric.

Definition We say that an operator H with dense domain L in a Hilbert space $\mathcal{H}$ is symmetric if for all $f, g \in L$ we have

$$\langle Hf, g\rangle = \langle f, Hg\rangle.$$

Considering again the operator $Hf = -f''$ on $L^2(a,b)$, choose either of the domains of Example 1.1.1. By use of the identity

$$\int_a^b (f''\overline{g} - f\overline{g''})\mathrm{d}x = [f'\overline{g} - f\overline{g'}]_a^b$$

we see that both H_D and H_N are symmetric.

Lemma 1.1.4 *Every symmetric operator H is closable and its closure is also symmetric.*

Proof Let $\mathcal{D}$ be the set of $f \in \mathcal{H}$ for which there exist $f_n \in \mathrm{Dom}(H)$ and $g \in \mathcal{H}$ such that $\lim_{n\to\infty} f_n = f$ and $\lim_{n\to\infty} Hf_n = g$. It is easy to see that $\mathcal{D}$ is a linear subspace of $\mathcal{H}$ containing $\mathrm{Dom}(H)$. If $h \in \mathrm{Dom}(H)$ then

$$\langle g, h\rangle = \lim_{n\to\infty} \langle Hf_n, h\rangle = \lim_{n\to\infty} \langle f_n, Hh\rangle = \langle f, Hh\rangle.$$

Now g is uniquely determined by the functional $h \to \langle g, h\rangle$ on $\mathrm{Dom}(H)$ because $\mathrm{Dom}(H)$ is dense in $\mathcal{H}$ by hypothesis. Therefore g is uniquely determined by f. If we define $\overline{H}f = g$ then it follows that $\overline{H}$ is linear on its domain $\mathcal{D}$. Moreover, the graph of $\overline{H}$ is the closure of the graph of H. If $h_n \in \text{Dom}(H)$, $\lim_{n\to\infty} h_n = h \in \mathcal{D}$ and $\lim_{n\to\infty} Hh_n = k$ then we have already shown that

$$\langle \overline{H}f, h_n\rangle = \langle f, Hh_n\rangle.$$

In the limit as $n \to \infty$ we get

$$\langle \overline{H}f, h\rangle = \langle f, k\rangle = \langle f, \overline{H}h\rangle$$

which establishes that $\overline{H}$ is symmetric. □

The above lemma enables us to concentrate henceforth on closed operators. The above definitions and Lemma 1.1.4 allow us to be a little careless in distinguishing between a closable operator and its closure and we shall often take advantage of this.

1.2 Self-adjointness

There is a difference between symmetry and self-adjointness for an unbounded operator A on a Hilbert space $\mathcal{H}$, which does not correspond to anything in the theory of bounded linear operators. At first this seems to be an annoying technicality, but in fact it is of profound importance. The condition of self-adjointness is much more demanding and difficult to verify, but unless it is met one cannot apply the very powerful machinery of spectral theory. In the context of differential operators, the issue is whether one has fully specified the boundary conditions appropriate to the particular differential operator one is studying.

Definition If A is a linear operator on a Hilbert space $\mathcal{H}$ then the adjoint operator A^* is determined by the condition that

$$\langle Af, g\rangle = \langle f, A^*g\rangle$$

for all $f \in \text{Dom}(A)$ and $g \in \text{Dom}(A^*)$. The domain of A^* is defined to be the set $\mathcal{D}$ of all $g \in \mathcal{H}$ for which there exists $k \in \mathcal{H}$ such that

$$\langle Af, g\rangle = \langle f, k\rangle$$

for all $f \in \text{Dom}(A)$. After showing that k is unique and we will put $A^*g = k$.

Lemma 1.2.1 *If A is a closed linear operator with dense domain then the adjoint A^* is also a closed linear operator with dense domain.*

Proof If $g \in \mathscr{D}$ and k, k' are two elements of $\mathscr{H}$ such that

$$\langle Af, g \rangle = \langle f, k \rangle = \langle f, k' \rangle$$

for all $f \in \mathrm{Dom}(A)$ then $\langle f, k - k' \rangle = 0$ for all such f. The density of $\mathrm{Dom}(A)$ implies that $k = k'$. It is thus permissible to define A^* on $\mathscr{D}$ by $A^* g = k$. If L is the graph of A^* then $(g, k) \in L$ if and only if (g, k) is orthogonal to

$$M := \{(Af, -f) \in \mathscr{H} \times \mathscr{H} : f \in \mathrm{Dom}(A)\}.$$

The orthogonal complement $M^\perp$ of the linear subspace M must be a closed linear subspace, so A^* is a closed linear operator.

It only remains to prove that A^* is densely defined. If $h \in \mathscr{H}$ satisfies $\langle h, g \rangle = 0$ for all $g \in \mathscr{D}$, then $(h, 0)$ is orthogonal to L. But $L^\perp = M^{\perp\perp} = \overline{M}$. Hence there exists a sequence $\{f_n\}_{n=1}^\infty \in \mathrm{Dom}(A)$ such that $\lim_{n\to\infty} f_n = 0$ and $\lim_{n\to\infty} Af_n = h$. But A is assumed to be closed, so $h = A0 = 0$. Thus $\mathscr{D}^\perp = 0$, and $\mathscr{D}$ is a dense linear subspace of $\mathscr{H}$. □

If H is a symmetric operator then it is easy to see that the adjoint H^* is an extension of H. We say that H is self-adjoint if H is symmetric and $\mathrm{Dom}(H) = \mathrm{Dom}(H^*)$. This is equivalent to requiring that $H = H^*$, and implies that H is closed. We say that H is essentially self-adjoint if it is symmetric and its closure is self-adjoint. Our next lemma gives a method of proving essential self-adjointness, but it is only useful for simple operators whose eigenvectors can be determined explicitly. In the proof of the lemma we assume that $\mathscr{H}$ is separable, or equivalently that it has a countable complete orthonormal set. This is valid in all applications to differential operators, and the interested reader can no doubt provide the necessary modification to non-separable Hilbert spaces. We shall make the same assumption at many other places in the book without comment.

Lemma 1.2.2 *Let H be a symmetric operator on $\mathscr{H}$ with domain L, and let $\{f_n\}_{n=1}^\infty$ be a complete orthonormal set in $\mathscr{H}$. If each f_n lies in L and there exist $\lambda_n \in \mathbf{R}$ such that $Hf_n = \lambda_n f_n$ for every n, then H is essentially self-adjoint. Moreover, the spectrum of $\overline{H}$ is the closure in $\mathbf{R}$ of the set of all λ_n.*

Proof If $f = \sum_{n=1}^\infty \alpha_n f_n$ lies in L and

$$g := Hf = \sum_{n=1}^\infty \beta_n f_n,$$

then

$$\beta_m = \langle g, f_m \rangle = \langle Hf, f_m \rangle = \langle f, Hf_m \rangle = \lambda_m \langle f, f_m \rangle = \lambda_m \alpha_m.$$

The requirements that $f, g \in \mathscr{H}$ force

$$\sum_{n=1}^{\infty} |\alpha_n|^2 < \infty \quad , \quad \sum_{n=1}^{\infty} |\beta_n|^2 < \infty$$

and hence

$$\sum_{n=1}^{\infty} (1 + \lambda_n^2)|\alpha_n|^2 < \infty.$$

We now define an operator $\tilde{H}$ as follows. Let $\tilde{L}$ be the set of all $f \in \mathscr{H}$ of the form $f = \sum_{n=1}^{\infty} \alpha_n f_n$ where

$$\sum_{n=1}^{\infty} (1 + \lambda_n^2)|\alpha_n|^2 < \infty,$$

and for such f define

$$\tilde{H}f := \sum_{n=1}^{\infty} \alpha_n \lambda_n f_n.$$

It is clear that $\tilde{H}$ is an extension of H. We first determine its spectrum.

Let S be the closure of the set $\{\lambda_n : 1 \le n < \infty\}$. Each λ_n is an eigenvalue of $\tilde{H}$, and $\mathrm{Spec}(\tilde{H})$ is closed, so $S \subseteq \mathrm{Spec}(\tilde{H})$. If $z \notin S$ then the operator A defined on $\mathscr{H}$ by

$$A\left(\sum_{n=1}^{\infty} \alpha_n f_n\right) := \sum_{n=1}^{\infty} \alpha_n (z - \lambda_n)^{-1} f_n$$

is bounded and one-one. By expanding everything in terms of the complete orthonormal set $\{f_n\}_{n=1}^{\infty}$ one can check that its range is precisely $\tilde{L}$ and that $(z - \tilde{H})Af = f$ for all $f \in \mathscr{H}$. Thus $z \notin \mathrm{Spec}(\tilde{H})$ and $A = (z - \tilde{H})^{-1}$. The above together imply that $S = \mathrm{Spec}(\tilde{H})$.

We claim that $\tilde{H}$ is equal to the closure of H. Since $\mathrm{Spec}(\tilde{H})$ is not equal to $\mathbf{C}$, Lemma 1.1.2 implies that $\tilde{H}$ is a closed operator. If $g := \sum_{n=1}^{\infty} \alpha_n f_n \in \mathrm{Dom}(\tilde{H})$ and we put $g_m := \sum_{n=1}^{m} \alpha_n f_n$ then $\lim_{m\to\infty} g_m = g$ and

$$\lim_{m\to\infty} \{Hg_m\} = \lim_{m\to\infty} \left\{\sum_{n=1}^{m} \alpha_n \lambda_n f_n\right\} = \sum_{n=1}^{\infty} \alpha_n \lambda_n f_n = \tilde{H}g.$$

This establishes the stated claim.

We finally prove that $\tilde{H}$ is self-adjoint. If $f \in \mathrm{Dom}(\tilde{H}^*)$ and $\tilde{H}^* f = k$, then

$$\begin{aligned}\langle k, f_n\rangle &= \langle \tilde{H}^* f, f_n\rangle \\ &= \langle f, \tilde{H} f_n\rangle \\ &= \lambda_n \langle f, f_n\rangle.\end{aligned}$$

If $f = \sum_{n=1}^{\infty} \alpha_n f_n$ the above implies that $k = \sum_{n=1}^{\infty} \lambda_n \alpha_n f_n$. It follows that $f \in \mathrm{Dom}(\tilde{H})$, and hence that $\tilde{H}^* = \tilde{H}$. □

Example 1.2.3 The operator H_D defined in Example 1.1.1 is essentially self-adjoint on its domain L_D. To see this we first define the orthonormal sequence of functions

$$f_n(x) := \left(\frac{2}{b-a}\right)^{1/2} \sin\left\{\frac{n\pi(x-a)}{(b-a)}\right\},$$

where $n \in \mathbf{N}$. The fact that these are eigenfunctions with eigenvalues $n^2\pi^2/(b-a)^2$ is easy to verify. The harder fact needed to apply Lemma 1.2.2 is that $\{f_n\}_{n=1}^{\infty}$ is a complete orthonormal set in $L^2(a,b)$; this is a standard result of Fourier analysis, but was proved long after the classic paper of Fourier (1822). A similar argument applies to H_N. □

Example 1.2.4 Another example of a similar type arises in connection with Legendre's equation

$$-\frac{\mathrm{d}}{\mathrm{d}x}\left\{(1-x^2)\frac{\mathrm{d}f}{\mathrm{d}x}\right\} = \lambda f.$$

We show that the symmetric operator H on $L^2(-1,1)$ defined by

$$Hf := -\frac{\mathrm{d}}{\mathrm{d}x}\left\{(1-x^2)\frac{\mathrm{d}f}{\mathrm{d}x}\right\}$$

is essentially self-adjoint on the domain of all twice continuously differentiable functions on $[-1,1]$. It is elementary but tedious to check that the Legendre polynomials $P_n(x)$ defined for $n \geq 0$ by

$$P_n(x) := \frac{1}{2^n n!}\frac{\mathrm{d}^n}{\mathrm{d}x^n}\left\{(x^2-1)^n\right\}$$

satisfy Legendre's equation with eigenvalues $\lambda_n = n(n+1)$. A further direct computation, using integration by parts repeatedly, establishes that the functions

$$f_n(x) := \left(\frac{2n+1}{2}\right)^{1/2} P_n(x)$$

form an orthonormal set in $L^2(-1,1)$. The linear span of the f_n coincides with the set of all polynomials on $[-1,1]$; this set is uniformly dense in $C[-1,1]$ by the Stone–Weierstrass theorem, and hence is norm dense in $L^2(-1,1)$. This implies that $\{f_n\}_{n=1}^{\infty}$ is a complete orthonormal set in $L^2(-1,1)$, which is what is needed to apply Lemma 1.2.2. We thus finally see that H is essentially self-adjoint on its domain. □

Example 1.2.5 One of the most important differential operators, the Laplacian $-\Delta$, is also one of the oldest. It was studied by Laplace (1749–1827) in connection with the theory of gravitation. If Ω is a region in $\mathbf{R}^N$ and C_0^∞ is the space of smooth (i.e. infinitely differentiable) functions on the closure $\overline{\Omega}$ of Ω which vanish on its boundary, then the symmetry of $-\Delta$, that is the identity

$$\int_\Omega \{(\Delta f)\overline{g} - f(\overline{\Delta g})\}\mathrm{d}^N x = 0$$

for all $f, g \in C_0^\infty$, follows by the use of a theorem of Green (1793–1841). The study of the spectral theory of this operator will occupy a substantial part of this book. □

Many symmetric differential operators are not essentially self-adjoint, and it is important to know whether they have any self-adjoint extensions. An abstract description of all symmetric extensions of a given symmetric operator H may be given as follows. If $f \in \mathrm{Dom}(H)$, then

$$\|(H+i)f\|^2 = \|Hf\|^2 + \|f\|^2 = \|(H-i)f\|^2. \qquad (1.2.1)$$

Therefore there exists an isometric linear operator $U = (H-i)(H+i)^{-1}$ mapping $\mathrm{Ran}(H+i)$ one-one onto $\mathrm{Ran}(H-i)$. We call this the Cayley transform of H.

Lemma 1.2.6 *There exists a one-one correspondence between symmetric extensions of H and isometric extensions of its Cayley transform U.*

Proof If $\tilde{H}$ is a symmetric extension of H then its Cayley transform $\tilde{U}$ is an isometric extension of U. Conversely let $\tilde{U}$ be an isometric extension of U mapping the linear subspace $M^+ \supseteq \mathrm{Ran}(H+i)$ onto the linear subspace $M^- \supseteq \mathrm{Ran}(H-i)$. Let $f \in M^+$ satisfy $\tilde{U}f = f$ and let $g = (H+i)h$ where $h \in \mathrm{Dom}(H)$. Then

$$\begin{aligned} 2i\langle f,h\rangle &= \langle f,(H-i)h-(H+i)h\rangle \\ &= \langle f, Ug-g\rangle \\ &= \langle \tilde{U}f,\tilde{U}g\rangle - \langle f,g\rangle \\ &= 0. \end{aligned}$$

Since $\mathrm{Dom}(H)$ is dense in $\mathcal{H}$ it follows that $f = 0$. Therefore the operator $(\tilde{U}-1)$ is one-one.

We now define the operator $\tilde{H}$ on $\mathcal{D} = (\tilde{U}-1)M^+$ by

$$(\tilde{H}+i)^{-1} := \tfrac{1}{2}i(\tilde{U}-1).$$

Equivalently we have $\tilde{H}f = \frac{1}{2}(\tilde{U}+1)g$ if $f = \frac{1}{2}i(\tilde{U}-1)g$ and $g \in M^+$. Indexing f and g in an obvious way we have

$$\begin{aligned} \langle \tilde{H}f_1, f_2\rangle &= \langle \tfrac{1}{2}(\tilde{U}+1)g_1, \tfrac{1}{2}i(\tilde{U}-1)g_2\rangle \\ &= -\tfrac{1}{4}i\{-\langle \tilde{U}g_1,g_2\rangle + \langle g_1,\tilde{U}g_2\rangle\} \\ &= \langle \tfrac{1}{2}i(\tilde{U}-1)g_1, \tfrac{1}{2}(\tilde{U}+1)g_2\rangle \\ &= \langle f_1, \tilde{H}f_2\rangle. \end{aligned}$$

Thus $\tilde{H}$ is a symmetric extension of H and $\tilde{U}$ is its Cayley transform. □

One of the best-known conditions for essential self-adjointness is in terms of deficiency subspaces and deficiency indices. We define the deficiency indices of a symmetric operator H to be the dimensions (possibly infinite) of the deficiency subspaces:

$$\begin{aligned} L^{\pm} &:= \{f \in \mathrm{Dom}(H^*) : H^*f = \pm if\} \\ &= \{f \in \mathcal{H} : \langle Hh, f\rangle = \mp i\langle h,f\rangle \quad \text{for all} \quad h \in \mathrm{Dom}(H)\}. \end{aligned}$$

One of the consequences of the following theorem is the fact that the closure of the Cayley transform is unitary if and only if the deficiency indices are both zero.

Theorem 1.2.7 *If H is a symmetric operator on $\mathcal{H}$ then there exist self-adjoint extensions of H if and only if the deficiency indices are equal. Moreover, the following conditions are equivalent:*

(1) *H is essentially self-adjoint,*
(2) *The deficiency indices of H are both zero,*
(3) *H has exactly one self-adjoint extension.*

Proof We first note that the deficiency subspaces of H and of its closure are equal, so there is no loss in assuming that H is closed.

(1) $\Rightarrow$ (2) If H is self-adjoint and $f \in L^+$ then $Hf = H^* f = if$. Therefore

$$i\langle f,f\rangle = \langle Hf,f\rangle = \langle f,Hf\rangle = -i\langle f,f\rangle$$

from which we see that $f = 0$. The proof that $L^- = 0$ is similar.

(2) $\Rightarrow$ (1) The operator $(H+i)$ maps $\mathrm{Dom}(H)$ one-one onto a subspace $M^+ := \mathrm{Ran}(H+i)$ of $\mathscr{H}$ and the inverse operator $(H+i)^{-1}$ defined on M^+ is a contraction by (1.2.1). Since H is closed we see that the inverse operator above is closed and bounded; hence the subspace M^+ must be closed. The orthogonal complement of M^+ is L^+, which is zero by hypothesis, and this implies that $M^+ = \mathscr{H}$. Therefore for any $f \in \mathrm{Dom}(H^*)$ there exists $g \in \mathrm{Dom}(H)$ such that $(H+i)g = (H^*+i)f$. But H^* is an extension of H so $(H^*+i)(f-g) = 0$. Therefore $(f-g) \in L^- = 0$. This proves that $f \in \mathrm{Dom}(H)$ and hence that $H^* = H$.

(2) $\Rightarrow$ (3) Lemma 1.2.6 shows even more, that H has no proper symmetric extension.

(3) $\Rightarrow$ (1) If H has a self-adjoint extension K then the deficiency indices of K must be zero by (2). By Lemma 1.2.6 the Cayley transform V of K is a unitary extension of the Cayley transform U of H. This extension maps L^+ one-one onto L^-, so the deficiency indices of H must be equal. Assuming for the moment that the deficiency indices are not zero, we obtain a different unitary extension of U for every isometric map of L^+ onto L^-, so there are infinitely many different unitary extensions. An application of Lemma 1.2.6 now shows that there are many different symmetric extensions with zero deficiency indices; these are all self-adjoint extensions of H by parts (1) and (2) of this theorem. The contradiction establishes that the deficiency indices are zero, and hence that H is essentially self-adjoint. □

One of the applications of the above theorem is the following. We leave the reader to formulate an abstract version of the condition below.

Lemma 1.2.8 *Let X be a measurable subset of $\mathbf{R}^N$ and let $\mathscr{H} = L^2(X)$. Let H be a symmetric operator on $\mathscr{H}$, which is real in the sense that if f lies in* Dom(H) *then its complex conjugate $\overline{f}$ also lies in* Dom(H) *and*

$$\overline{H(f)} = H(\overline{f}).$$

Then H has at least one self-adjoint extension.

Proof One only needs to observe that H^* is real and hence that the conjugation operator $f \to \overline{f}$ maps L^+ one-one onto L^-. These subspaces must therefore have the same dimension. □

It is certainly not the case that such operators are normally essentially self-adjoint. For example if H is defined on $L^2(a, b)$ as in Example 1.1.1 but with domain

$$\begin{aligned} L :&= L_N \cap L_D \\ &= \{f \in C^2[a, b] : f(a) = f'(a) = f(b) = f'(b) = 0\}, \end{aligned}$$

then H is real but has at least two different self-adjoint extensions, namely $\overline{H_D}$ and $\overline{H_N}$.

By identifying the deficiency subspaces explicitly, one can sometimes use Lemma 1.2.6 to classify all self-adjoint extensions of a given symmetric operator. However, this is usually only possible for ordinary differential operators, for which the deficiency subspaces are finite-dimensional.

Example 1.2.9 Let H be the symmetric differential operator on $L^2(0, \infty)$ defined by

$$Hf := if'$$

with domain the space $C_c^\infty(0, \infty)$ of smooth functions with compact supports within $(0, \infty)$. Anticipating some ideas from the theory of distributions in Sections 3.4 and 6.1, one may show that $H^*f = \pm if$ if and only if $f, f' \in L^2$ and $f' = \pm f$. This forces $f(x) = ce^{\pm x}$. Therefore $\dim(L^+) = 0$ and $\dim(L^-) = 1$, and H has no self-adjoint extensions. If, however, we consider the 'same' operator acting on $L^2(0, 1)$ with domain $C_c^\infty(0, 1)$, then H has deficiency indices $1, 1$ and an application of Theorem 1.2.7 shows that H has infinitely many self-adjoint extensions. □

The following result is the key to all further work on the spectral theory of self-adjoint operators. Its importance is that it will allow us to transfer attention from the unbounded operator H to the bounded resolvent operators.

Theorem 1.2.10 *The spectrum of any self-adjoint operator H is real and non-empty. If $z \notin \mathbf{R}$ then*

$$\|(z - H)^{-1}\| \le |\mathrm{Im}\, z|^{-1}. \tag{1.2.2}$$

Moreover,

$$(\bar{z} - H)^{-1} = ((z - H)^{-1})^*. \tag{1.2.3}$$

Proof We put $K = (H - x)/y$ where $z = x + iy$ and $y \neq 0$. It may be seen that $K = K^*$. The proof of Theorem 1.2.7 now establishes that $\pm i \notin \mathrm{Spec}(K)$ and $\|(K \pm i)^{-1}\| \leq 1$. These facts imply that $z \notin \mathrm{Spec}(H)$ and that (1.2.2) holds. If $f_1, f_2 \in \mathrm{Dom}(H)$ then

$$\langle (H - z)f_1, f_2 \rangle = \langle f_1, (H - \bar{z})f_2 \rangle.$$

Putting $g_1 = (H - z)f_1$ and $g_2 = (H - \bar{z})f_2$ and rewriting the last equation in terms of g_1 and g_2 yields (1.2.3).

Finally suppose that the spectrum of H is empty. Then for any $\phi, \psi \in \mathcal{H}$ the complex-valued function

$$f(z) := \langle (z - H)^{-1}\phi, \psi \rangle$$

is analytic on $\mathbf{C}$ and vanishes as $|z| \to \infty$. Liouville's theorem now implies that f is identically zero. Since ϕ, ψ are arbitrary we conclude that $(z - H)^{-1}$ is identically zero. The falsity of this conclusion shows that the spectrum of H is non-empty. □

1.3 Multiplication operators

We consider an example of a self-adjoint operator which appears to have nothing to do with the theory of partial differential operators, but which wi, be central to the description and application of the spectral theorem. We describe the example less generally than is possible, but nevertheless more generally than we shall need.

Let E be a Borel subset of $\mathbf{R}^N$ and let μ be a (non-negative countably additive Borel) measure which is finite on every bounded Borel subset of $\mathbf{R}^N$. We define $\mathcal{H} := L^2(E, \mathrm{d}\mu)$ to be the space of all measurable functions $f : E \to \mathbf{C}$ such that

$$\|f\| := \{\int_E |f(x)|^2 \mathrm{d}\mu\}^{1/2} < \infty.$$

Subject to identifying two functions on E if they are equal almost everywhere, $\mathcal{H}$ is a Hilbert space.

Let $a : E \to \mathbf{R}$ be a measurable function such that the restriction of a to any bounded subset of E is a bounded function. Let $\mathcal{D}$ be the set of

$f \in \mathcal{H}$ such that

$$\int_E \{1 + a(x)^2\}|f(x)|^2 \mathrm{d}\mu < \infty.$$

We now define the operator A with domain $\mathcal{D}$ by

$$(Af)(x) := a(x)f(x).$$

Lemma 1.3.1 *The operator A is self-adjoint on the domain $\mathcal{D}$. If L_c^2 is the set of functions $f \in \mathcal{H}$ which vanish outside some bounded subset of E, then A is essentially self-adjoint on L_c^2.*

Proof We first observe that the operator A is symmetric on its domain. Let $z \notin \mathbf{R}$ and consider the bounded operator R_z on $\mathcal{H}$ defined by

$$(R_z f)(x) := \{z - a(x)\}^{-1} f(x).$$

If $f \in \mathcal{D}$ then it is elementary that

$$R_z(z - A)f = f,$$

while if $f \in \mathcal{H}$ it is also elementary that $R_z f \in \mathcal{D}$ and that

$$(z - A)R_z f = f.$$

Therefore $z \notin \mathrm{Spec}(A)$ and R_z is the resolvent of A. Lemma 1.1.2 implies that A is closed. The deficiency subspaces of A are both zero, so A is self-adjoint by Theorem 1.2.7.

Now let $f \in \mathcal{D}$ and for any $n \in \mathbf{N}$ define

$$f_n(x) := \begin{cases} f(x) & \text{if } x \in E \text{ and } |x| < n \\ 0 & \text{otherwise.} \end{cases}$$

Since $|f_n(x)| \le |f(x)|$ for all $x \in E$ we see that $f_n(x) \in L_c^2 \subseteq \mathcal{D}$ for all n. Moreover,

$$\lim_{n\to\infty} \|f_n - f\| = 0 \quad , \quad \lim_{n\to\infty} \|Af_n - Af\| = 0$$

by the use of the dominated convergence theorem in both cases. Therefore A is the closure of its restriction to L_c^2. □

The essential range of a function, defined below, is the closest one can get to the set of values which the function takes, if the function is only defined up to possible alteration on a set of measure zero.

Lemma 1.3.2 *The spectrum of A equals the essential range of a, that is the set of all $\lambda \in \mathbf{R}$ such that*

$$\mu\{x : |a(x) - \lambda| < \varepsilon\} > 0$$

for all $\varepsilon > 0$. If $\lambda \notin \mathrm{Spec}(A)$ then

$$\{(\lambda - A)^{-1} f\}(x) = \{\lambda - a(x)\}^{-1} f(x)$$

for all $x \in E$ and $f \in \mathscr{H}$, and

$$\|(\lambda - A)^{-1}\| = [\mathrm{dist}(\lambda, \mathrm{Spec}(A))]^{-1}.$$

Proof If λ does not lie in the essential range of a then the function $r_\lambda(x) := \{\lambda - a(x)\}^{-1}$ is bounded outside some null set, and therefore determines a bounded multiplication operator R_λ on L^2. It is easy to check that the range of this operator equals $\mathrm{Dom}(A)$ and that $(\lambda - A)R_\lambda = 1$ as an operator identity on L^2; the proof that $R_\lambda(\lambda - A) = 1$ is similar. Therefore $\lambda \notin \mathrm{Spec}(A)$. Conversely if λ lies in the essential range of a then the sets

$$S_m := \{x : |\lambda - a(x)| < 2^{-m}\}$$

have non-zero measures for all m; if the measure of S_m is infinite then we replace S_m by a subset of positive but finite measure. If ϕ_m is the characteristic function of S_m then $0 \neq \phi_m \in L^2$ and

$$\|(\lambda - A)\phi_m\| \leq 2^{-m}\|\phi_m\|.$$

It follows that the operator $(\lambda - A)$ cannot have a bounded inverse, so $\lambda \in \mathrm{Spec}(A)$.

The last statement of the theorem follows from the identity

$$\|(\lambda - A)^{-1}\| = \text{ess-sup}\{|\lambda - a(x)|^{-1} : x \in E\}$$

whose proof is similar to that of the first statement. □

1.4 Relatively bounded perturbations

Let H be a self-adjoint operator with dense domain $\mathscr{D}$ in a Hilbert space $\mathscr{H}$. Let A be a symmetric operator whose domain contains $\mathscr{D}$. We say that A has relative bound $\alpha \geq 0$ with respect to H if there exists $c < \infty$ such that

$$\|Af\| \leq \alpha\|Hf\| + c\|f\| \tag{1.4.1}$$

for all $f \in \mathcal{D}$. This condition is satisfied with $\alpha = 0$ if A is a bounded linear operator with domain $\mathcal{H}$. We define $K : \mathcal{D} \to \mathcal{H}$ by $Kf := Hf + Af$.

Lemma 1.4.1 *If $0 \le \alpha < 1$, then the operator K is closed and symmetric on $\mathcal{D}$. Moreover, $\|A(H + i\lambda)^{-1}\| < 1$ for large enough real λ.*

Proof The symmetry of K is elementary. Suppose $f_n \in \mathcal{D}$, $\lim_{n\to\infty} \|f_n - f\| = 0$ and $\lim_{n\to\infty} \|Kf_n - g\| = 0$. The inequality

$$\begin{aligned}\|Hf_n - Hf_m\| &\le \|Kf_n - Kf_m\| + \|Af_n - Af_m\| \\ &\le \|Kf_n - Kf_m\| + \alpha\|Hf_n - Hf_m\| + c\|f_n - f_m\|\end{aligned}$$

implies

$$\|Hf_n - Hf_m\| \le (1-\alpha)^{-1}\|Kf_n - Kf_m\| + c(1-\alpha)^{-1}\|f_n - f_m\|.$$

It follows from the closedness of H that $f \in \mathcal{D}$ and $\lim_{n\to\infty} Hf_n = Hf$. An application of (1.4.1) now implies that $\lim_{n\to\infty} Af_n = Af$. These two formulae together imply that $\lim_{n\to\infty} Kf_n = Kf$, so K is closed.

Secondly let $g \in \mathcal{H}$ and put $f := (H \pm i\lambda)^{-1}g$. If $0 \le \alpha < \beta < 1$ then

$$\begin{aligned}\|Af\|^2 &\le (\alpha\|Hf\| + c\|f\|)^2 \\ &\le \beta^2(\|Hf\|^2 + \lambda^2\|f\|^2) \\ &= \beta^2\|(H \pm i\lambda)f\|^2 \\ &= \beta^2\|g\|^2,\end{aligned}$$

provided $\lambda > 0$ is large enough. This implies that $\|A(H \pm i\lambda)^{-1}\| \le \beta$, as required. □

Theorem 1.4.2 *Let H be self-adjoint and let A be symmetric with relative bound less than 1 with respect to H. Then the operator $K := H + A$ is self-adjoint with* Dom(K) = Dom(H).

Proof If $\lambda \in \mathbf{R}$ is large enough then the series

$$R_\lambda := (H - i\lambda)^{-1} \sum_{n=0}^{\infty} \{-A(H - i\lambda)^{-1}\}^n$$

is norm convergent. It is immediate that R_λ has range equal to $\mathcal{D}$, and a straightforward calculation establishes that

$$(H + A - i\lambda)R_\lambda f = f$$

for all $f \in \mathscr{H}$. It follows that

$$\mathrm{Ran}(H + A - i\lambda) = \mathscr{H}.$$

After repeating the above with λ replaced by $-\lambda$, an application of Theorem 1.2.7 now proves that $\lambda^{-1}(H + A)$ and hence $(H + A)$ are self-adjoint. □

Theorem 1.4.2, proved by Rellich in 1939, is one of the most famous methods of establishing that a given symmetric operator K is self-adjoint. Its systematic application to non-relativistic quantum theory is due to many people, but particularly Kato (1966). Its use depends upon being able to express K as a perturbation of another operator H, which is already known to be self-adjoint. Its weakness is that for many problems, including those of the type studied in Chapters 6 and 7, no such comparison operators can be found. We shall therefore make rather little use of the theorem. There is a quadratic form analogue, Corollary 4.4.3, which is considerably more useful for our purposes.

Exercises

1.1 Determine all the eigenvalues and eigenfunctions of the operators H_D and H_N of Example 1.1.1.

1.2 Use Lemma 1.2.2 to prove that the operators of Example 1.1.1 are essentially self-adjoint on their domains.

1.3 Let H_p be defined on the dense subspace

$$L_p := \{f \in C^2[-\pi, \pi] : f(-\pi) = f(\pi) \text{ and } f'(-\pi) = f'(\pi)\}$$

of $L^2(-\pi, \pi)$ by $H_p f := -f''$. Prove that H_p is essentially self-adjoint and find its spectrum.

1.4 Let $\theta \in \mathbf{R}$ and let H_θ be defined on the dense domain

$$\mathscr{D}_\theta := \{f \in C^\infty[0, 1] : f(1) = \mathrm{e}^{i\theta} f(0)\}$$

in $L^2(0, 1)$ by

$$H_\theta f(x) := i\frac{\mathrm{d}f}{\mathrm{d}x}.$$

Find all of the eigenvalues and eigenfunctions of H_θ. Prove that H_θ is essentially self-adjoint on $\mathscr{D}_\theta$.

1.5 The operator A is defined on the dense domain $C_c^\infty(a,b)$ of $L^2(a,b)$ by the formula

$$Af := \sum_{r=0}^{n} a_r \frac{\mathrm{d}^r f}{\mathrm{d}x^r}.$$

Prove that $C^\infty[a,b] \subseteq \mathrm{Dom}(A^*)$ and that

$$A^*g = \sum_{r=0}^{n} (-1)^r \overline{a_r} \frac{\mathrm{d}^r g}{\mathrm{d}x^r}$$

for all $g \in C^\infty[a,b]$.

1.6 (For those who need something to occupy a long winter weekend.) Let H be the operator $Hf := -f''$ with dense domain $C_c^\infty(a,b) \subseteq L^2(a,b)$. Assuming that the deficiency subspaces $L^\pm$ are contained in $C^\infty[a,b]$, prove that H has deficiency indices $(2,2)$. Use Lemma 1.2.6 to find all of the self-adjoint extensions of H, expressing your answer in terms of boundary conditions at a and b.

1.7 Prove that a unitary operator U acting on a Hilbert space $\mathcal{H}$ is the Cayley transform of some self-adjoint operator if and only if 1 is not an eigenvalue of U.

1.8 If A is a closed, densely defined operator on a Hilbert space $\mathcal{H}$, prove that $A^{**} = A$.

1.9 The operator H is defined on a domain of sufficiently smooth functions in $L^2(\mathbf{R}^2)$ by the formula

$$Hf := -a\frac{\partial^2 f}{\partial x^2} - 2b\frac{\partial^2 f}{\partial x \partial y} - c\frac{\partial^2 f}{\partial y^2},$$

where a, b and c are real constants. Find necessary and sufficient conditions on these constants for there to exist a linear change of the variables $(x,y) \in \mathbf{R}^2$ under which the operator transforms into $\tilde{H}f := -\Delta f$.

1.10 Let A and B be two operators with the same domain $\mathcal{D}$ in a Hilbert space H. Put

$$\alpha := \inf\{a \in \mathbf{R}^+ : \|Af\| \le a\|Bf\| + b\|f\| \text{ for some } b < \infty \text{ and all } f \in \mathcal{D}\}$$

$$\beta := \inf\{a \in \mathbf{R}^+ : \|Af\|^2 \le a^2\|Bf\|^2 + b^2\|f\|^2 \text{ for some } b < \infty \text{ and all } f \in \mathcal{D}\}.$$

Prove that $\alpha = \beta$.

1.11 Let V be a bounded measurable real-valued function on $[-\pi, \pi]$, and define H on the subspace L_p of Exercise 1.3 by

$$Hf := H_p f + Vf.$$

Use Theorem 1.4.2 to prove that H is essentially self-adjoint on L_p.

1.12 (Hard) Define B on L_p by

$$Bf(x) := \frac{\mathrm{d}f}{\mathrm{d}x}.$$

Use Fourier series methods to prove that for all $\varepsilon > 0$ there exists $c_\varepsilon < \infty$ such that

$$\|Bf\|_2 \leq \varepsilon\|H_p f\|_2 + c_\varepsilon\|f\|_2$$

for all $f \in L_p$.

1.13 (Hard) Define H on L_p by

$$Hf(x) := -\gamma(x)\frac{\mathrm{d}^2 f}{\mathrm{d}x^2} - \gamma'(x)\frac{\mathrm{d}f}{\mathrm{d}x},$$

where γ is a continuously differentiable periodic function on $[-\pi, \pi]$ which satisfies $0 < c \leq \gamma(x) \leq 1$ for all $x \in [-\pi, \pi]$. Use Theorem 1.4.2 to compare H with the operator H_p of Exercise 1.9, and prove that H is essentially self-adjoint on L_p.

2
The spectral theorem

2.1 Introduction

The general spectral theorem for self-adjoint operators was proved independently by Stone and von Neumann during the period 1929–1932. There have been several other proofs since that time, the most popular being that based upon Gelfand's theory of commutative Banach algebras. There are also several different ways of stating the spectral theorem, one in terms of a functional calculus, one using a family of spectral projections and one in terms of a measure-theoretic representation formula; all of these have advantages.

In this chapter we describe an approach to the spectral theorem which originates from a paper written in 1989 by Helffer and Sjöstrand. The approach is very explicit and has proved of great value in doing computations in n-body scattering theory. It has been rewritten in an axiomatic form applicable to suitable operators on Banach spaces in Davies (1994), but here we shall present only the simpler theory for self-adjoint operators. Another advantage of this approach is that it uses techniques of a type which are useful in other parts of the theory of partial differential equations, rather than abstract functional analysis. The material in this chapter is of fundamental importance to later work, but the reader may choose to defer reading the proofs to a later stage. The key results are Theorems 2.3.1, 2.5.1 and 2.5.3.

Our proof of the spectral theorem is based upon presenting an explicit formula for $f(H)$ as an integral over resolvents, for a fairly large class of functions f. The work then consists of showing that this formula has all of the properties required of a functional calculus. Only at the end do we extend the functional calculus to the space $C_0(\mathbf{R})$ of all continuous functions on $\mathbf{R}$ which vanish at $\pm\infty$.

Throughout this chapter we take H to be an arbitrary bounded or unbounded self-adjoint operator on an abstract separable Hilbert space $\mathcal{H}$. The critical property of H which we need was proved in Theorem 1.2.10 and is the bound

$$\|(z-H)^{-1}\| \le |\mathrm{Im}(z)|^{-1} \tag{2.1.1}$$

valid for all $z \notin \mathbf{R}$.

If $z \in \mathbf{C}$, then we write $\langle z \rangle := (1+|z|^2)^{1/2}$; some properties of this function are listed in the exercises. If $\beta \in \mathbf{R}$, then we define S^β to be the set of smooth functions $f : \mathbf{R} \to \mathbf{C}$ such that

$$|f^{(n)}(x)| := \left|\frac{\mathrm{d}^n f}{\mathrm{d}x^n}\right| \le c_n \langle x \rangle^{\beta - n}$$

for some $c_n < \infty$, all $x \in \mathbf{R}$ and all integers $n \ge 0$. We then consider the class of slowly decreasing smooth functions f on $\mathbf{R}$; more precisely we define $\mathcal{A}$ by

$$\mathcal{A} := \bigcup_{\beta<0} S^\beta. \tag{2.1.2}$$

The important features of $\mathcal{A}$ are that it is an algebra under pointwise multiplication, and that it contains every rational function whose denominator does not vanish on $\mathbf{R}$ and is of higher degree than its numerator. (See Exercise 2.1; we advise readers to check for themselves any other assertions for which no proof is written out.) We define the norms

$$\|f\|_n := \sum_{r=0}^{n} \int_{-\infty}^{\infty} |f^{(r)}(x)| \langle x \rangle^{r-1} \mathrm{d}x \tag{2.1.3}$$

on $\mathcal{A}$ for all integers $n \ge 1$. We shall actually only need these norms for $n \le 3$, but higher values of n are needed for some applications. We comment that the finiteness of $\|f\|_n$ implies that $f' \in L^1(\mathbf{R})$ and hence that $f \in C_0(\mathbf{R})$. Norm convergence of a sequence of functions with respect to $\|\cdot\|_n$ implies uniform convergence in $C_0(\mathbf{R})$.

In the next section we will give an explicit definition of the operator $f(H)$ for all $f \in \mathcal{A}$. We will show that the map $f \to f(H)$ from $\mathcal{A}$ to the algebra $\mathcal{L}(\mathcal{H})$ of all bounded operators on $\mathcal{H}$ has the following properties:

(1) The map $f \to f(H)$ is linear and multiplicative (i.e. is an algebra homomorphism).
(2) We have $\overline{f}(H) = f(H)^*$ for all $f \in \mathcal{A}$.
(3) We have $\|f(H)\| \le \|f\|_\infty$ for all $f \in \mathcal{A}$.

(4) If $w \notin \mathbf{R}$ and $r_w(s) := (w-s)^{-1}$, then $r_w(H) = (w-H)^{-1}$.
(5) If $f \in C_c^\infty(\mathbf{R})$ has support disjoint from $\mathrm{Spec}(H)$, then $f(H) = 0$.

Under the above conditions we say that $f \to f(H)$ is a functional calculus. The next section is devoted to the proof of all of the above assertions, and following sections follow up their implications.

2.2 The Helffer–Sjöstrand formula

The starting point for our analysis contains the only novel idea in the whole theory. If $f : \mathbf{R} \to \mathbf{C}$ is a smooth (i.e. infinitely differentiable) function we define a smooth extension $\tilde{f} : \mathbf{C} \to \mathbf{C}$ by

$$\tilde{f}(z) := \Big\{\sum_{r=0}^{n} f^{(r)}(x)\,(iy)^r/r!\Big\}\sigma(x,y), \tag{2.2.1}$$

where $n \geq 1$ and

$$\sigma(x,y) := \tau(y/\langle x\rangle)$$

for some smooth function $\tau(s)$, defined on $\mathbf{R}$, which equals 1 if $|s| < 1$ and equals 0 if $|s| > 2$. The exact choice of n and τ turns out not to be important. The extension $\tilde{f}$ is smooth as a function of (x,y) but is not analytic.

We shall make constant use of the formula

$$\begin{aligned}\frac{\partial \tilde{f}(z)}{\partial \bar{z}} &:= \tfrac{1}{2}\Big\{\frac{\partial \tilde{f}}{\partial x} + i\frac{\partial \tilde{f}}{\partial y}\Big\}\\ &= \tfrac{1}{2}\Big\{\sum_{r=0}^{n} f^{(r)}(x)\,(iy)^r/r!\Big\}\{\sigma_x + i\sigma_y\}\\ &\quad + \tfrac{1}{2}f^{(n+1)}(x)\,(iy)^n(n!)^{-1}\sigma.\end{aligned} \tag{2.2.2}$$

It follows from this expression that

$$\left|\frac{\partial \tilde{f}}{\partial \bar{z}}\right| = O(|y|^n)$$

as $y \to 0$ for each $x \in \mathbf{R}$. In particular

$$\frac{\partial \tilde{f}}{\partial \bar{z}} = 0$$

for every $z \in \mathbf{R}$, which is why we call $\tilde{f}$ an almost analytic extension of f.

Helffer and Sjöstrand used the functional calculus, which they assumed already known, to prove that the bounded operator $f(H)$ on $\mathcal{H}$ is given for every $f \in \mathcal{A}$ by the equation

$$f(H) := -\frac{1}{\pi}\int_{\mathbf{C}} \frac{\partial \tilde{f}(z)}{\partial \overline{z}}(z-H)^{-1}\mathrm{d}x\mathrm{d}y. \qquad (2.2.3)$$

We, however, use this formula to define the operator $f(H)$, and then verify that $f \to f(H)$ has all the properties needed of a functional calculus. Our method of proof uses a few results from analytic function theory for functions of a complex variable which are operator-valued rather than complex-valued. We reassure the reader that the proofs of these results are exactly the same as the standard proofs, apart from the occasional replacement of absolute value signs by norms.

Lemma 2.2.1 *The integral (2.2.3) is norm convergent and*

$$\|f(H)\| \le c_n \|f\|_{n+1}$$

for all $f \in \mathcal{A}$ and all $n \ge 1$, where the norms are those defined by (2.1.3).

Proof It follows from Lemma 1.1.2 that the integrand is norm continuous for $z \notin \mathbf{R}$. This enables us to define the integral over any compact subset of $\mathbf{C}\backslash\mathbf{R}$ as a norm limit of approximating sums. The convergence in norm of these integrals as the compact set increases to the whole of $\mathbf{C}\backslash\mathbf{R}$ depends upon obtaining strong enough bounds on the norm of the integrand over $\mathbf{C}\backslash\mathbf{R}$. If

$$U := \{(x,y) : \langle x\rangle < |y| < 2\langle x\rangle\} \quad , \quad V := \{(x,y) : 0 < |y| < 2\langle x\rangle\},$$

then

$$|\sigma_x(z) + i\sigma_y(z)| \le c\langle x\rangle^{-1}\chi_U(z)$$

for all $z \in \mathbf{C}$; see Exercise 2.3. Therefore the norm of the integrand of (2.2.3) is dominated by

$$c\sum_{r=0}^{n} |f^{(r)}(x)|\langle x\rangle^{r-2}\chi_U(x,y) + c|f^{(n+1)}(x)|\,|y|^{n-1}\chi_V(x,y).$$

If $n \ge 1$ then integrating this with respect to y yields the bound

$$\|f(H)\| \le c\|f\|_{n+1}. \qquad \square$$

Although we are interested in functions lying in $\mathcal{A}$, several of the proofs below are made easier by restricting attention to functions in $C_c^\infty(\mathbf{R})$. This does not involve any loss of generality because of our next lemma.

Lemma 2.2.2 *The space $C_c^\infty(\mathbf{R})$ is dense in $\mathcal{A}$ for each of the norms $\| \ \|_{n+1}$.*

Proof Suppose that $f \in S^\beta$ for some $\beta < 0$. Given $\phi \in C_c^\infty$ such that $\phi(s) = 1$ if $|s| < 1$ and $\phi(s) = 0$ if $|s| > 2$, put $\phi_m(s) = \phi(s/m)$ and $f_m = \phi_m f$. If $n \geq 1$, then

$$\|f - f_m\|_{n+1} = \sum_{r=0}^{n+1} \int_{-\infty}^{\infty} \left| \frac{\mathrm{d}^r}{\mathrm{d}x^r} \left\{ f(x)(1 - \phi_m(x)) \right\} \right| \langle x \rangle^{r-1} \mathrm{d}x.$$

We expand this using Leibnitz' formula and the bound

$$\left| \frac{\mathrm{d}^s}{\mathrm{d}x^s} \phi_m(x) \right| \leq c_s m^{-s} \chi_m(x) \leq c_s' \langle x \rangle^{-s} \chi_m(x),$$

valid for $s \geq 1$, where χ_m is the characteristic function of $\{x : |x| > m\}$. This yields

$$\|f - f_m\|_{n+1} \leq c \sum_{r=0}^{n+1} \int_{|x|>m} \left| \frac{\mathrm{d}^r f}{\mathrm{d}x^r} \right| \langle x \rangle^{r-1} \mathrm{d}x$$

which converges to zero as $m \to \infty$. □

Our next lemma is crucial for establishing that the operator $f(H)$ is independent of the choices of n and σ.

Lemma 2.2.3 *If $F \in C_c^\infty(\mathbf{C})$ and $F(z) = O(y^2)$ as $y \to 0$ for each $x \in \mathbf{R}$, then*

$$-\frac{1}{\pi} \int_{\mathbf{C}} \frac{\partial F}{\partial \bar{z}} (z - H)^{-1} \mathrm{d}x\mathrm{d}y = 0. \tag{2.2.4}$$

Proof Let F have support in $\{z : |x| < N \text{ and } |y| < N\}$ and define Ω_δ for small $\delta > 0$ to be the region $\{z : |x| < N \text{ and } \delta < |y| < N\}$. If A denotes the left-hand side of (2.2.4) then an application of Stokes' formula yields

$$\begin{aligned} A &= -\lim_{\delta \to 0} \frac{1}{\pi} \int_{\Omega_\delta} \frac{\partial F}{\partial \bar{z}} (z - H)^{-1} \mathrm{d}x\mathrm{d}y \\ &= \lim_{\delta \to 0} \frac{i}{2\pi} \int_{\partial \Omega_\delta} F(z)(z - H)^{-1} \mathrm{d}z. \end{aligned}$$

The last integral is a sum of integrals over eight line segments, six of which equal 0 since they do not meet the support of F. Therefore

$$\|A\| \le \lim_{\delta\to 0}\frac{1}{2\pi}\int_{-N}^{N}\{|F(x+i\delta)|+|F(x-i\delta)|\}\delta^{-1}\mathrm{d}x = 0. \qquad \square$$

Lemma 2.2.4 *If $f \in \mathcal{A}$ and $n \ge 1$ then the operator $f(H)$ is independent of the choices of σ and of n.*

Proof The norm density result of Lemma 2.2.2 implies that it is sufficient to prove this if $f \in C_c^\infty$. For the purpose of this proof we use the more complicated notation $\tilde{f}_{\sigma,n}$ to stand for the right-hand side of (2.2.1).

If $f \in C_c^\infty(\mathbf{R})$ and σ_1, σ_2 are two cut-off functions then $\tilde{f}_{\sigma_1,n} - \tilde{f}_{\sigma_2,n}$ equals 0 for small $|y|$ so Lemma 2.2.3 proves that $\tilde{f}_{\sigma,n}(H)$ is independent of σ. If $m > n \ge 1$, then $\tilde{f}_{\sigma,n} - \tilde{f}_{\sigma,m} = O(y^2)$ for small $|y|$ so Lemma 2.2.3 proves that $\tilde{f}_{\sigma,n}(H)$ is independent of n. $\square$

Lemma 2.2.5 *If $f \in C_c^\infty(\mathbf{R})$ has support disjoint from the spectrum of H then $f(H) = 0$.*

Proof There exist a finite number of closed curves $\{\gamma_r\}_{r=1}^n$ which together enclose a region U which contains the support of $\tilde{f}$ and is disjoint from the spectrum of H. An application of Stokes' theorem yields

$$\begin{aligned} f(H) &= -\frac{1}{\pi}\int_U \frac{\partial\tilde{f}(z)}{\partial\bar{z}}(z-H)^{-1}\mathrm{d}x\mathrm{d}y \\ &= \frac{i}{2\pi}\sum_{r=1}^{n}\int_{\gamma_r}\tilde{f}(z)(z-H)^{-1}\mathrm{d}z, \end{aligned}$$

which vanishes since $\tilde{f}$ equals 0 on each of the γ_r. $\square$

Lemma 2.2.6 *We have*

$$(fg)(H) = f(H)g(H)$$

for all $f, g \in \mathcal{A}$.

Proof We first assume that f and g lie in $C_c^\infty(\mathbf{R})$. Let $K := \mathrm{supp}(\tilde{f})$ and $L := \mathrm{supp}(\tilde{g})$ so that K, L are compact subsets of $\mathbf{C}$ and

$$f(H)g(H) = \frac{1}{\pi^2}\int_{K\times L}\frac{\partial\tilde{f}}{\partial\bar{z}}\frac{\partial\tilde{g}}{\partial\bar{w}}(z-H)^{-1}(w-H)^{-1}\mathrm{d}x\mathrm{d}y\mathrm{d}u\mathrm{d}v.$$

The resolvent identity

$$(z-H)^{-1}(w-H)^{-1}=(z-w)^{-1}(w-H)^{-1}-(z-w)^{-1}(z-H)^{-1}$$

implies that we may expand this as the sum of two integrals, each of which is norm absolutely integrable. Each of these may be simplified by means of the following formulae, proved using Cauchy's theorem:

$$-\frac{1}{\pi}\int_K \frac{\partial\tilde{f}}{\partial\bar{z}}(z-w)^{-1}\mathrm{d}x\mathrm{d}y=\tilde{f}(w),$$
$$\frac{1}{\pi}\int_L \frac{\partial\tilde{g}}{\partial\bar{w}}(z-w)^{-1}\mathrm{d}u\mathrm{d}v=\tilde{g}(z).$$

This leads to the identity

$$\begin{aligned}f(H)g(H)&=-\frac{1}{\pi}\int_{K\cup L}\left\{\tilde{f}(z)\frac{\partial\tilde{g}}{\partial\bar{z}}+\frac{\partial\tilde{f}}{\partial\bar{z}}\tilde{g}(z)\right\}(z-H)^{-1}\mathrm{d}x\mathrm{d}y\\&=-\frac{1}{\pi}\int_{K\cup L}\frac{\partial(\tilde{f}\tilde{g})}{\partial\bar{z}}(z-H)^{-1}\mathrm{d}x\mathrm{d}y.\end{aligned}$$

In order to prove that this equals $(fg)(H)$ we need to show that

$$-\frac{1}{\pi}\int_{\mathbf{C}}\frac{\partial k}{\partial\bar{z}}(z-H)^{-1}\mathrm{d}x\mathrm{d}y=0,$$

where

$$k(z):=\tilde{f}(z)\tilde{g}(z)-(fg)\tilde{\ }(z).$$

Since k is of compact support and vanishes to high order on the real axis, this is a simple application of Lemma 2.2.3.

Now suppose that $f,g\in S^{\beta}$ for some $\beta<0$. Given $\phi\in C_c^{\infty}$ such that $\phi(s)=1$ if $|s|<1$ and $\phi(s)=0$ if $|s|>2$, define ϕ_m by $\phi_m(s)=\phi(s/m)$. If $f_m=\phi_m f$, $g_m=\phi_m g$, and $h_m=\phi_m^2 fg$, then it follows from the above that

$$h_m(H)=f_m(H)g_m(H).$$

The proof is completed by using Lemmas 2.2.1 and 2.2.2. □

Up to the present point it is entirely possible that $f(H)=0$ for all $f\in\mathcal{A}$. Proving that this is not the case is the hardest part of our approach to the spectral theorem.

Lemma 2.2.7 *If $w\notin\mathbf{R}$ and $r_w(z):=(w-z)^{-1}$ for all $z\neq w$ then $r_w\in\mathcal{A}$ and*

$$r_w(H)=(w-H)^{-1}.$$

Proof According to Lemma 2.2.4 we may choose σ and n arbitrarily within certain limits without affecting the operators $f(H)$. In this proof we put $n := 1$. We assume for definiteness that $\mathrm{Im}(w) > 0$ and put

$$\sigma(z) := \tau(\lambda|y|/\langle x\rangle),$$

where $\lambda \geq 1$ is large enough to ensure that $w \notin \mathrm{supp}(\sigma)$. For large $m > 0$ define

$$\Omega_m := \{(x, y) : |x| < m \text{ and } m^{-1}\langle x\rangle < |y| < 2m\}.$$

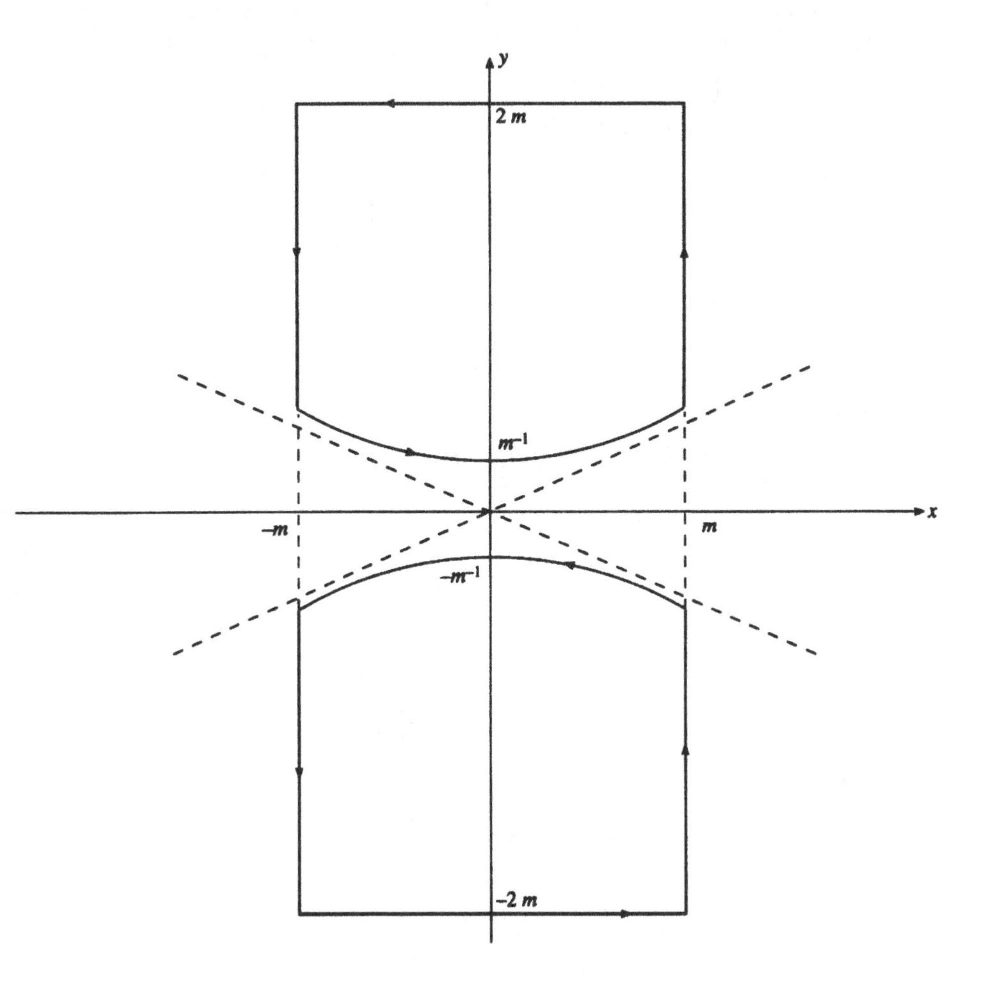

The boundary of Ω_m consists of two closed curves, both traversed in the anticlockwise direction. An application of Stokes' theorem yields

$$r_w(H) = -\frac{1}{\pi}\lim_{m\to\infty}\int_{\Omega_m}\frac{\partial\tilde{r}_w}{\partial\bar{z}}(z-H)^{-1}\mathrm{d}x\mathrm{d}y$$
$$= \frac{i}{2\pi}\lim_{m\to\infty}\int_{\partial\Omega_m}\tilde{r}_w(z)(z-H)^{-1}\mathrm{d}z.$$

We next show that

$$\lim_{m\to\infty}\|\int_{\partial\Omega_m}\{r_w(z)-\tilde{r}_w(z)\}(z-H)^{-1}\mathrm{d}z\| = 0.$$

This involves independent estimates for each of the six straight lines and two curves in $\partial\Omega_m$. If z lies on one of the vertical lines γ then $\langle z\rangle = O(m)$ and $\langle x\rangle = O(m)$. An application of Taylor's theorem with $n = 1$ to the function $y \to r_w(x+iy)$ shows that

$$|r_w(z)-\tilde{r}_w(z)| \le \big|\big(1-\sigma(z)\big)r_w(z)\big|$$
$$+\,\sigma(z)|r_w(z)-r_w(x)-r_w'(x)iy|$$
$$\le c\chi_W(z)\langle x\rangle^{-1} + c\frac{|y|^2}{\langle x\rangle^3},$$

where

$$\chi_W(z) := \begin{cases} 1 & \text{if } \langle x\rangle < \lambda|y|, \\ 0 & \text{otherwise.}\end{cases}$$

Therefore

$$\|\int_\gamma\{r_w(z)-\tilde{r}_w(z)\}(z-H)^{-1}\mathrm{d}z\|$$
$$\le c\int_{\lambda^{-1}\langle m\rangle}^{2m}\frac{\mathrm{d}y}{my} + c\int_{\lambda^{-1}\langle m\rangle}^{2m}\frac{y\mathrm{d}y}{m^3}$$
$$= O(m^{-1}).$$

If γ' is one of the curves $y = \pm\langle x\rangle/m$, then $\sigma(z) = 1$ for all $z \in \gamma'$ and

$$|r_w(z)-\tilde{r}_w(z)| \le c\frac{|y|^2}{\langle x\rangle^3}.$$

Therefore

$$\|\int_{\gamma'}\{r_w(z)-\tilde{r}_w(z)\}(z-H)^{-1}\mathrm{d}z\|$$
$$\le c\int_{\gamma'}\frac{|y|^2}{\langle x\rangle^3}\frac{1}{|y|}|\mathrm{d}z|$$
$$= c\,m^{-1}\int_{\gamma'}\langle x\rangle^{-2}|\mathrm{d}z|$$
$$= O(m^{-1}).$$

Finally if z lies on one of the horizontal straight lines γ'' then $\sigma(z) = 0$, while $|y|$ and $\langle z \rangle$ are both of order m. Therefore

$$\| \int_{\gamma''} \{r_w(z) - \tilde{r}_w(z)\}(z - H)^{-1} \mathrm{d}z \| \le c \int_{\gamma''} \frac{\mathrm{d}x}{m^2} = O(m^{-1}).$$

Combining these bounds we obtain

$$r_w(H) = \frac{i}{2\pi} \lim_{m \to \infty} \int_{\partial \Omega_m} r_w(z)(z - H)^{-1} \mathrm{d}z.$$

The integrand is holomorphic on and inside the part of $\partial\Omega_m$ in the lower half-plane, so the contribution of that integral is zero by Cauchy's theorem. The integrand is meromorphic in the upper half-plane with a single pole at $z = w$. Therefore

$$\begin{aligned} r_w(H) &= -\mathrm{Res}_{z=w}\{r_w(z)(z - H)^{-1}\} \\ &= (w - H)^{-1}. \end{aligned}$$

□

The last lemma in the series is the only one which makes essential use of the facts that $\mathscr{H}$ is a Hilbert space and H is self-adjoint.

Lemma 2.2.8 *We have*

$$\bar{f}(H) = f(H)^*$$

and

$$\|f(H)\| \le \|f\|_\infty$$

for all $f \in \mathscr{A}$.

Proof The first statement is an immediate consequence of the identity $(\bar{z} - H)^{-1} = [(z - H)^{-1}]^*$. To prove the second let $f \in \mathscr{A}$ and let $c > \|f\|_\infty$. If we define

$$g(s) := c - \{c^2 - |f(s)|^2\}^{1/2},$$

then a direct computation (Exercise 2.5) shows that $g \in \mathscr{A}$. From the identity

$$f\bar{f} - 2cg + g^2 = 0$$

in the algebra $\mathscr{A}$, we deduce that

$$f(H)^* f(H) - cg(H) - cg(H)^* + g(H)^* g(H) = 0$$

or equivalently

$$f(H)^* f(H) + \{c - g(H)\}^* \{c - g(H)\} = c^2.$$

If $\psi \in \mathcal{H}$, then it follows that

$$\|f(H)\psi\|^2 \leq \|f(H)\psi\|^2 + \|\{(c - g(H)\}\psi\|^2$$
$$= c^2\|\psi\|^2.$$

This implies the second statement of the lemma. □

2.3 The first spectral theorem

We are now in a position to collect together all of the lemmas of the last section into a major theorem. In this theorem we assume that H is a given self-adjoint operator on a Hilbert space $\mathcal{H}$, and that $C_0(\mathbf{R})$ denotes the space of all continuous functions on $\mathbf{R}$ which vanish at $\pm\infty$, with the supremum norm. The following is the first of three versions of the spectral theorem.

Theorem 2.3.1 *There exists a unique linear map $f \to f(H)$ from $C_0(\mathbf{R})$ to $\mathcal{L}(\mathcal{H})$ such that:*

(1) The map $f \to f(H)$ is multiplicative (i.e. is an algebra homomorphism).
(2) We have $\overline{f}(H) = f(H)^*$ for all $f \in C_0(\mathbf{R})$.
(3) We have $\|f(H)\| \leq \|f\|_\infty$ for all $f \in C_0(\mathbf{R})$.
(4) If $w \notin \mathbf{R}$ and $r_w(s) := (w - s)^{-1}$ then $r_w(H) = (w - H)^{-1}$.
(5) If $f \in C_0(\mathbf{R})$ has support disjoint from $\mathrm{Spec}(H)$ then $f(H) = 0$.

Proof If we replace $C_0(\mathbf{R})$ by $\mathcal{A}$ then the existence of such a map $f \to f(H)$ is established by the series of lemmas in the last section. Now $\mathcal{A}$ is a dense linear subspace of $C_0(\mathbf{R})$ by the Stone–Weierstrass theorem, and the existence of an extension to $C_0(\mathbf{R})$ follows from (3). The fact that the extension has the required properties is routine.

It remains only to prove the uniqueness. If we have two such maps $f \to f(H)$ then by (4) they coincide on the linear subspace

$$\left\{ \sum_{i=1}^{n} \lambda_i r_{w_i} : \lambda_i \in \mathbf{C} \;\&\; w_i \notin \mathbf{R} \right\}.$$

This subspace is dense in $C_0(\mathbf{R})$ by the Stone–Weierstrass theorem. The fact that the two maps $f \to f(H)$ coincide on $C_0(\mathbf{R})$ now follows from (3). □

The above theorem will be refined below in some very important respects. In Theorem 2.5.1 we use it to give a representation of any self-adjoint operator as a multiplication operator on a suitable L^2 space. This representation can then be used to extend the definition of the operator $f(H)$ to bounded measurable functions f; see Theorem 2.5.3. These three theorems together constitute the spectral theorems, and will be used repeatedly throughout this text.

2.4 Invariant and cyclic subspaces

Let H be a self-adjoint operator acting on a Hilbert space $\mathcal{H}$ and let L be a closed linear subspace of $\mathcal{H}$. Then we say that L is invariant if $(z-H)^{-1}L \subseteq L$ for all $z \notin \mathbf{R}$. If H is bounded we show below that this is equivalent to the condition that $HL \subseteq L$. Invariant subspaces are of such importance to the study of self-adjoint operators that there has been much effort to try to extend their use to general bounded operators; however, it is still unknown whether every bounded linear operator on a Hilbert space must have at least one invariant subspace other than 0 and $\mathcal{H}$.

If $\lambda \in \mathbf{R}$ then the eigenspace associated to H is defined to be

$$L_\lambda := \{f \in \mathrm{Dom}(H) : Hf = \lambda f\}$$

and is non-zero provided λ is an eigenvalue. Using the fact that H is a closed operator, it is easy to see that L_λ is a closed linear subspace and that it is invariant. The dimension of L_λ is called the multiplicity of λ.

Example 2.4.1 If H is an ordinary differential operator of order n acting on an interval $I \subseteq \mathbf{R}$, then every eigenvalue has multiplicity at most n. This is because the initial value problem for ordinary differential equations states that there are at most n independent solutions of the eigenvalue equation. The boundary conditions will usually further reduce the dimension of the eigenspace. □

Lemma 2.4.2 *If L is an invariant subspace for the self-adjoint operator H, then its orthogonal complement $L^\perp$ is also invariant. Moreover, $f(H)L \subseteq L$ for all $f \in C_0(\mathbf{R})$.*

Proof If $\psi \in L^\perp$, then the invariance of L implies that

$$\langle (z-H)^{-1}\psi, \phi\rangle = \langle \psi, (\overline{z}-H)^{-1}\phi\rangle = 0$$

for all $\phi \in L$. Therefore $(z-H)^{-1}\psi \in L^\perp$ and $L^\perp$ is invariant. If

L is invariant then the Helffer–Sjöstrand formula (2.2.3) implies that $f(H)L \subseteq L$ for all $f \in \mathscr{A}$. The same holds for all $f \in C_0(\mathbf{R})$ by a density argument. □

We next consider one particularly important type of invariant subspace. If A is a bounded self-adjoint operator on $\mathscr{H}$, we say that L is a cyclic subspace for A with cyclic vector $v \in \mathscr{H}$ if L is the norm closure of the linear span of all vectors $p(A)v$, where p is any polynomial in one variable. If H is an unbounded self-adjoint operator, we require L to be the norm closure of the linear span of all vectors $(z-H)^{-1}v$ where $z \notin \mathbf{R}$. We write

$$L = \overline{\text{lin}}\{(z-H)^{-1}v : z \notin \mathbf{R}\}$$

and call L the cyclic subspace generated by v.

Lemma 2.4.3 *If H is a bounded self-adjoint operator then the two definitions of invariant subspace and of cyclic subspace are equivalent. For any self-adjoint operator H on $\mathscr{H}$, bounded or not, there exists a sequence of orthogonal cyclic subspaces L_n of $\mathscr{H}$ with cyclic vectors v_n such that $\mathscr{H}$ is the closure of the algebraic sum $\sum_{n=1}^{\infty} \oplus L_n$.*

Proof We indicate the two notions of invariant and cyclic by superscripts ^{b} and ^{g} to denote bounded and general. Let the closed subspace L be invariant^b, that is $HL \subseteq L$, and let $|z| > \|H\|$. Then by using the norm convergent power series

$$(z-H)^{-1} = \sum_{n=0}^{\infty} z^{-n-1}H^n,$$

we see that $(z-H)^{-1}L \subseteq L$. If

$$S := \{w \in \mathbf{C}\backslash\text{Spec}(H) : (w-H)^{-1}L \subseteq L\},$$

then by using the resolvent expansion (1.1.6) we can extend the set S until it includes every point in $\mathbf{C}\backslash\text{Spec}(H)$; in topological terms S is open and closed in the connected set $\mathbf{C}\backslash\text{Spec}(H)$ and therefore coincides with it. Therefore L is invariant^g.

Conversely let L be invariant^g, that is $(z-H)^{-1}L \subseteq L$ for all $z \in \mathbf{C}\backslash\mathbf{R}$. Let $f \in L$ and let $Hf = g + h$ where $g \in L$ and $h \in L^\perp$. An application of Lemma 2.4.2 implies that $(z-H)^{-1}h \in L^\perp$. Since $f \in L$ and $zf - g \in L$, we see that

$$(z-H)^{-1}h = (z-H)^{-1}(Hf-g) = (z-H)^{-1}(zf-g) - f \in L.$$

Therefore $(z-H)^{-1}h \in L \cap L^{\perp} = \{0\}$. But $(z-H)^{-1}$ is one-one, so $h = 0$. This implies that $HL \subseteq L$ and that L is invariant^b.

Now it is easy to see that a closed subspace L is cyclic^b (resp. cyclic^g) with cyclic vector v if and only if it is the smallest invariant^b (resp. invariant^g) subspace which contains v. This establishes that the two definitions of cyclic subspace are equivalent.

For the second part of the proof we assume that $\mathcal{H}$ contains a countable dense subset $\{f_n\}_{n=1}^{\infty}$: we invite the reader to modify the statement and proof of the lemma to the case of non-separable Hilbert spaces by using Zorn's lemma. Let L_1 be the cyclic subspace generated by f_1. Inductively, let us assume that we are given orthogonal cyclic subspaces $L_1, L_2, ..., L_n$. Let $m(n)$ be the smallest integer for which $f_{m(n)} \notin L_1 \oplus ... \oplus L_n$. Let $g_{m(n)}$ be the component of $f_{m(n)}$ orthogonal to $L_1 \oplus ... \oplus L_n$ and let L_{n+1} be the cyclic subspace generated by $g_{m(n)}$. It is easy to show that L_{n+1} is orthogonal to L_r for all $r \leq n$ and that $f_{m(n)} \in L_1 \oplus ... \oplus L_{n+1}$.

Inductively there are two possibilities. If no $m(n)$ exists at some stage, then $\mathcal{H}$ equals the finite sum of the cyclic subspaces obtained up to that stage. If the induction can be continued without end then the algebraic sum $\sum_{n=1}^{\infty} \oplus L_n$ contains all f_n and hence is dense in $\mathcal{H}$. In both cases the proof is complete. □

If H_n denotes the restriction of H to L_n, this lemma enables us to reduce all spectral questions concerning H to questions about H_n. Its main fault from the abstract point of view is that the direct sum decomposition of $\mathcal{H}$ into cyclic subspaces is not canonical, but this is not often important.

The theory of cyclic subspaces is important even in the numerical analysis of matrices. Let H be a self-adjoint $n \times n$ matrix, regarded as an operator on $\mathbf{C}^n$. By representing H as a diagonal matrix one may see that there exists a vector v for which $\mathbf{C}^n$ is a cyclic subspace if and only if every eigenvalue of H has multiplicity one. If we think of the eigenvalues as varying randomly, then the coincidence of two eigenvalues is an exceptional event, and we conclude that $\mathbf{C}^n$ is 'typically' cyclic for a self-adjoint matrix H.

We say that an $n \times n$ matrix A is tridiagonal if $|i-j| > 1$ implies that $A_{ij} = 0$. The numerical analysis of such matrices is particularly simple because most of their coefficients vanish.

Lemma 2.4.4 *Every self-adjoint $n \times n$ matrix H is unitarily equivalent to a tri-diagonal matrix. The procedure for implementing this equivalence is fully constructive.*

Proof By Lemma 2.4.3 it is sufficient to prove this in the case when $\mathbf{C}^n$ has a cyclic vector v. Let $\{v_i\}_{i=1}^n$ be defined by $v_i := H^{i-1}v$. If their linear span does not equal $\mathbf{C}^n$ then they do not form a linearly independent set, and for some $r \le n$ there exists a relation

$$v_r = \sum_{i=1}^{r-1} \alpha_i v_i.$$

We deduce that

$$Hv_r = \sum_{i=1}^{r-1} \alpha_i v_{i+1} \subseteq M.$$

Therefore $M := \mathrm{lin}\{v_i : 1 \le i \le r-1\}$ is a proper invariant subspace of C^n containing v, and C^n cannot be cyclic.

We now let $\{e_i\}_{i=1}^n$ be the orthonormal basis of C^n obtained by applying the Gram–Schmidt orthogonalisation procedure to the v_i. For every $r \le n$ we have

$$M_r := \mathrm{lin}\{v_i : 1 \le i \le r\} = \mathrm{lin}\{e_i : 1 \le i \le r\}.$$

Now $HM_r \subseteq M_{r+1}$ so $\langle He_r, e_s\rangle = 0$ if $s > r+1$. Finally the fact that H is self-adjoint implies that $\langle He_r, e_s\rangle = 0$ if $s < r-1$, and hence that the matrix of H with respect to the basis $\{e_i\}_{i=1}^n$ is tridiagonal. □

2.5 The L^2 spectral representation

The following theorem gives an explicit representation of any self-adjoint operator as a multiplication operator on an L^2-space, as described in Section 1.3. Functions of the operator are then represented by other multiplication operators constructed in an obvious fashion. The spectrum of the operator is easily identified as are many other spectral invariants, in spite of the fact that the representation is not canonical. The theorem may be summarised in the statement that every self-adjoint operator is unitarily equivalent to a multiplication operator, which is both easy to remember and ideal for applications to the solution of other spectral problems.

Theorem 2.5.1 *Let H be a self-adjoint operator on a Hilbert space $\mathcal{H}$ with spectrum S. Then there exists a finite measure μ on $S \times \mathbf{N}$ and a unitary operator*

$$U : \mathcal{H} \to L^2 := L^2(S \times \mathbf{N}, \mathrm{d}\mu)$$

with the following properties. If $h : S \times \mathbf{N} \to \mathbf{R}$ is the function $h(s,n) = s$, then the element ξ of $\mathscr{H}$ lies in Dom(H) *if and only if $h \cdot U(\xi) \in L^2$. We have*

$$UHU^{-1}\psi = h\psi$$

for all $\psi \in U\{\mathrm{Dom}(H)\}$, and also

$$Uf(H)U^{-1}\psi = f(h)\psi$$

for all $f \in C_0(\mathbf{R})$ and $\psi \in L^2(S \times \mathbf{N}d\mu)$.

We will deduce this theorem from the following special case.

Theorem 2.5.2 *Let H be a self-adjoint operator on a Hilbert space $\mathscr{H}$ with spectrum S, and suppose that $\mathscr{H}$ has the cyclic vector v. Then there exists a finite measure μ on S and a unitary operator*

$$U : \mathscr{H} \to L^2 := L^2(S, d\mu)$$

with the following properties. If $h : S \to \mathbf{R}$ is the function $h(s) = s$, then the element ξ of $\mathscr{H}$ lies in Dom(H) *if and only if $h \cdot U(f) \in L^2$. Moreover,*

$$UHU^{-1}\psi = h\psi$$

for all $\psi \in U\{\mathrm{Dom}(H)\}$.

Proof Consider the linear functional $\phi : C_0(\mathbf{R}) \to \mathbf{C}$ defined by

$$\phi(f) := \langle f(H)v, v\rangle.$$

It follows from the properties of the functional calculus that $\phi(\overline{f}) = \overline{\phi(f)}$ for all f. If $0 \le f \in C_0(\mathbf{R})$, then putting $g = f^{1/2}$ we see by the functional calculus that

$$\phi(f) = \|g(H)v\|^2 \ge 0.$$

The Riesz representation theorem now implies that there exists a finite countably additive measure μ on $\mathbf{R}$ such that

$$\langle f(H)v, v\rangle = \int_{\mathbf{R}} f(x)d\mu(x)$$

for all $f \in C_0(\mathbf{R})$. If f has support disjoint from S, then $f(H) = 0$ by the spectral theorem, and this implies that the measure μ has support in S.

The linear map $T : C_0(\mathbf{R}) \to L^2$ defined by $Tf = f$ has the property that

$$\begin{aligned}\langle Tf, Tg\rangle &= \int_S f(x)\overline{g(x)}\mathrm{d}\mu(x)\\ &= \phi(f\overline{g})\\ &= \langle g(H)^* f(H)v, v\rangle\\ &= \langle f(H)v, g(H)v\rangle\end{aligned}$$

for all $f, g \in C_0(\mathbf{R})$. If $\mathcal{M}$ denotes the linear subspace

$$\{f(H)v \in \mathcal{H} : f \in C_0(\mathbf{R})\},$$

then the above identity yields the existence of an isometric linear map U from $\mathcal{M}$ onto $C_0(\mathbf{R}) \subseteq L^2$ satisfying

$$U\big(f(H)v\big) = f$$

for all $f \in C_0(\mathbf{R})$. Now $\mathcal{M}$ is dense in $\mathcal{H}$ because v is cyclic, and fundamental properties of measure theory imply that $C_0(\mathbf{R})$ is dense in L^2. Therefore U may be extended to a unitary map from $\mathcal{H}$ onto L^2.

Now let $f_i, f \in C_0(\mathbf{R})$ for $i = 1, 2$ and let $\psi_i = f_i(H)v \in \mathcal{H}$. Then

$$\begin{aligned}\langle f(H)\psi_1, \psi_2\rangle &= \langle (f f_1)(H)v, f_2(H)v\rangle\\ &= \int_S f f_1 \overline{f_2}\mathrm{d}\mu(x)\\ &= \langle f U\psi_1, U\psi_2\rangle,\end{aligned}$$

where f denotes the obvious multiplication operator. Putting $f := r_w$ as in Theorem 2.3.1 we deduce that

$$Ur_w(H)U^{-1}\xi = r_w\xi \tag{2.5.1}$$

for all $\xi \in L^2$ and all $w \notin \mathbf{R}$.

It is evident from (2.5.1) that U maps the range of $r_w(H)$ one-one onto the range of the multiplication operator r_w. In other words U maps $\mathrm{Dom}(H)$ one-one onto the set of $\phi \in L^2$ such that $x\phi(x) \in L^2$. If $\xi \in L^2$ and we put $\psi = r_w\xi$, then $\psi \in \mathrm{Dom}(h)$ and from

$$Hr_w(H)U^{-1}\xi = wr_w(H)U^{-1}\phi - U^{-1}\xi$$

we deduce that

$$\begin{aligned}UHU^{-1}\psi &= UHU^{-1}r_w\xi\\ &= UHr_w(H)U^{-1}\xi\end{aligned}$$

$$= wr_w\xi - \xi$$
$$= h\psi. \qquad \square$$

Proof of Theorem 2.5.1 We first use Lemma 2.4.3 to write $\mathcal{H}$ as a direct sum of cyclic subspaces L_n with cyclic vectors v_n. Without loss of generality we may assume that $\|v_n\| = 2^{-n}$ for all $n \in N$. By Theorem 2.5.2 there exist measures μ_n, which must have masses 2^{-2n}, and unitary operators $U_n : L_n \to L^2(S, \mathrm{d}\mu_n)$ such that the restriction H_n of H to L_n is unitarily equivalent to multiplication by x on $L^2(S, \mathrm{d}\mu_n)$. The theorem now follows by defining the measure μ on $S \times N$ to have restriction μ_n on each subset $S \times \{n\}$, and then combining all of the maps in the obvious manner. $\square$

Even though the representation of Theorem 2.5.1 is not canonical, we can use it to prove an extension of the spectral theorem. In this extension $\mathcal{B}$ denotes the algebra of bounded Borel measurable functions on $\mathbf{R}$. We say that $f_n \in \mathcal{B}$ increases monotonically to $f \in \mathcal{B}$ if $f_n(x)$ increases pointwise and monotonically to $f(x)$ for every $x \in \mathbf{R}$. This implies that the supremum norms

$$\|f_n\| := \sup\{|f_n(x)| : x \in \mathbf{R}\}$$

are uniformly bounded. Note that we do not identify two functions in $\mathcal{B}$ if they are equal almost everywhere with respect to Lebesgue measure, because Lebesgue measure has no particular status in this context.

Our next theorem makes use of the notion of strong convergence. If $\{A_n\}_{n=1}^{\infty}$ is a sequence of bounded operators on a Banach space $\mathcal{X}$ and A is a bounded operator we write $\text{s-lim}_{n\to\infty} A_n = A$ provided $\lim_{n\to\infty} \|A_n\phi - A\phi\| = 0$ for all $\phi \in \mathcal{X}$.

Theorem 2.5.3 *There exists a map $f \to f(H)$ from $\mathcal{B}$ to $\mathcal{L}(\mathcal{H})$ which extends the map of Theorem 2.3.1, and has the same properties, with the replacement of $C_0(\mathbf{R})$ by $\mathcal{B}$. The extension is unique subject to the further requirement that*

$$\operatorname*{s-lim}_{n\to\infty} f_n(H) = f(H)$$

whenever $f_n \in \mathcal{B}$ converges monotonically to $f \in \mathcal{B}$.

Proof We first prove existence. Using the unitary equivalence of

Theorem 2.5.1 to identify $\mathcal{H}$ with $L^2(S\times\mathbf{N}, \mathrm{d}\mu)$ and H with multiplication by h, we define $f(H)$ for all $f \in \mathcal{B}$ by

$$f(H)\phi(s,n) := f\{h(s,n)\}\phi(s,n).$$

The verification that the map $f \to f(H)$ has all of the required properties is routine, the limiting behaviour for monotone sequences f_n being a consequence of the dominated convergence theorem of measure theory.

The uniqueness statement of the theorem cannot use Theorem 2.5.1. Given two maps $f \to f(H)$ with the stated properties, let $\mathcal{C}$ be the set of all $f \in \mathcal{B}$ for which they coincide. Since both maps extend the functional calculus on $C_0(\mathbf{R})$ we know that $C_0(\mathbf{R}) \subseteq \mathcal{C}$. Moreover, $\mathcal{C}$ is closed under monotone limits because both maps $f \to f(H)$ are assumed to have the above continuity property. But $\mathcal{B}$ is the smallest class of functions containing $C_0(\mathbf{R})$ and closed under monotone limits, so $\mathcal{C} = \mathcal{B}$. □

Corollary 2.5.4 *The spectrum of H equals the essential range of h, even though the representation of Theorem 2.5.1 is not unique.*

Proof We first observe that because of the unitary equivalence, $\lambda \in \mathbf{C}$ lies in the spectrum of H if and only if it lies in the spectrum of the multiplication operator h on L^2. We may now quote Theorem 1.3.2. □

Our final version of the spectral theorem involves spectral projections. We start by summarising some standard facts. We define an (orthogonal) projection on a Hilbert space $\mathcal{H}$ to be a bounded linear operator P such that $P = P^* = P^2$. For every closed linear subspace $L \subseteq \mathcal{H}$ there exists a projection P on $\mathcal{H}$ with range equal to L and kernel equal to $L^\perp$. Moreover, every projection arises in this manner.

Theorem 2.5.5 *Let H be a self-adjoint operator on a Hilbert space $\mathcal{H}$ and let (a,b) be a bounded or unbounded open interval. Let f_n be an increasing sequence of non-negative continuous functions on $\mathbf{R}$ with supports in (a,b), which converge pointwise to the characteristic function χ of (a,b). Then the operators $f_n(H)$ converge strongly to a canonically determined projection $P_{(a,b)}$ which depends upon a and b but not upon the particular sequence f_n used. This projection commutes with H and vanishes if and only if $(a,b) \cap \operatorname{Spec}(H) = \emptyset$.*

Proof The existence of the strong limit of $f_n(H)$ is established by using Theorem 2.5.3, which also establishes that the limit equals $\chi(H)$, which depends only upon H, a and b. Since $\chi = \overline{\chi} = \chi^2$, we deduce from Theorem 2.5.3 that the limit P satisfies $P = P^* = P^2$. Using the notation of Theorem 2.5.1, we see that $P = 0$ if and only if $\{(s,n) : a < h(s,n) < b\}$ has zero μ-measure. This is equivalent to the condition that no point of (a,b) lies in the essential range of h, and hence by Corollary 2.5.4 to the condition that (a,b) does not meet the spectrum of H. □

If H is a self-adjoint operator then the projections $P_{(a,b)}$ are called spectral projections of H. Their range spaces $L_{(a,b)} := P_{(a,b)}\mathcal{H}$ are called spectral subspaces. In the notation of Theorem 2.5.1 they can be identified with the spaces $L^2(E_{(a,b)}, \mathrm{d}\mu)$, where

$$E_{(a,b)} := \{(s,n) : a < h(s,n) < b\}.$$

2.6 Norm resolvent convergence

There are several senses in which one can say that a sequence of self-adjoint operators on a Hilbert space $\mathcal{H}$ converges to a limit. We shall concentrate on the norm resolvent convergence of a sequence of self-adjoint operators $\{H_n\}_{n=1}^{\infty}$ to a limit H which is also known to be self-adjoint. We say that H_n converge in this sense to H if

$$\lim_{n\to\infty} \|(H_n + i)^{-1} - (H + i)^{-1}\| = 0.$$

Our first lemma establishes that the particular choice of i above has no significance.

Lemma 2.6.1 *If $z \in \mathbf{C}\backslash\mathbf{R}$ then*

$$\|(H_n - z)^{-1} - (H - z)^{-1}\| \le \frac{9\langle z\rangle^2}{|\mathrm{Im}(z)|^2} \|(H_n + i)^{-1} - (H + i)^{-1}\|.$$

Proof Our main task is to prove the operator identity

$$(H_n - z)^{-1} - (H - z)^{-1} = B\{(H_n + i)^{-1} - (H + i)^{-1}\}A, \tag{2.6.1}$$

where $A := (H + i)/(H - z)$ and $B := (H_n + i)/(H_n - z)$. The statement of the lemma follows from this by using bounds such as

$$\begin{aligned}\|A\| &= \|1 + (i + z)(H - z)^{-1}\| \\ &\le 1 + (1 + |z|)\,\|(H - z)^{-1}\| \\ &\le 1 + (1 + |z|)\,|\mathrm{Im}(z)|^{-1}.\end{aligned}$$

There is an easy proof of (2.6.1) which simply involves algebraic manipulations, forgetting that expressions such as $H_n(H-z)^{-1}$ may not make sense because $\mathrm{Ran}\{(H-z)^{-1}\}$ need not be contained in $\mathrm{Dom}(H_n)$. A correct proof is slightly more involved.

Using operator identities such as

$$1+z(H-z)^{-1}=H(H-z)^{-1}$$

we have

$$\begin{aligned}(H_n-z)^{-1}&-(H-z)^{-1}\\ &=\left[(H_n-z)^{-1}+z(H-z)^{-1}(H_n-z)^{-1}\right]\\ &\quad-\left[(H-z)^{-1}+z(H-z)^{-1}(H_n-z)^{-1}\right]\\ &=H(H-z)^{-1}.(H_n-z)^{-1}-(H-z)^{-1}.H_n(H_n-z)^{-1}\\ &=AH(H+i)^{-1}.(H_n+i)^{-1}B-A(H+i)^{-1}.H_n(H_n+i)^{-1}B\\ &=A\left[H(H+i)^{-1}.(H_n+i)^{-1}-(H+i)^{-1}.H_n(H_n+i)^{-1}\right]B\\ &=A\left[(H_n+i)^{-1}-(H+i)^{-1}\right]B.\end{aligned}$$

□

Theorem 2.6.2 *If H_n converge to H in the norm resolvent sense then*

$$\lim_{n\to\infty}\|f(H_n)-f(H)\|=0$$

for all $f\in C_0(\mathbf{R})$.

Proof If $f\in\mathcal{A}$ then an application of (2.2.3) implies that

$$\begin{aligned}\|f(H_n)-f(H)\|&\le\frac{1}{\pi}\int_{\mathbf{C}}\left|\frac{\partial\tilde{f}}{\partial\bar{z}}\right|\|(z-H_n)^{-1}-(z-H)^{-1}\|\mathrm{d}x\mathrm{d}y\\ &\le\frac{9}{\pi}\|(H_n+i)^{-1}-(H+i)^{-1}\|\int_{\mathbf{C}}\left|\frac{\partial\tilde{f}}{\partial\bar{z}}\right|\frac{\langle z\rangle^2}{|\mathrm{Im}(z)|^2}\mathrm{d}x\mathrm{d}y.\end{aligned}$$

The integral is finite if we take a suitable almost analytic extension of f, by the considerations of this chapter. The norm convergence now follows for $f\in\mathcal{A}$. The extension to all $f\in C_0(\mathbf{R})$ depends upon the use of Theorem 2.3.1 and the Stone–Weierstrass theorem. □

Corollary 2.6.3 *If H_n converges in the norm resolvent sense to H then the spectrum of H_n converges to the spectrum of H in the following sense. If $\lambda\in\mathbf{R}\backslash\mathrm{Spec}(H)$, then $\lambda\notin\mathrm{Spec}(H_n)$ for all large enough n. If $\lambda\in\mathrm{Spec}(H)\subseteq\mathbf{R}$, then there exist $\lambda_n\in\mathrm{Spec}(H_n)$ for which $\lim_{n\to\infty}\lambda_n=\lambda$.*

Proof Let $\lambda \in \mathbf{R}\backslash\mathrm{Spec}(H)$. There exists $f \in C_c^\infty(\mathbf{R})$ with $f(\lambda) = 1$ and $\mathrm{supp}(f) \cap \mathrm{Spec}(H) = \emptyset$. Since $f(H) = 0$ we see from the last theorem that $\lim_{n\to\infty} \|f(H_n)\| = 0$. An application of the Spectral Theorem implies that $\lambda \notin \mathrm{Spec}(H_n)$ as soon as $\|f(H_n)\| < 1$.

Conversely let $\lambda \in \mathrm{Spec}(H) \subseteq \mathbf{R}$. Let $\varepsilon > 0$ and let $f \in C_c^\infty$ satisfy $f(\lambda) = 1$ and $\mathrm{supp}(f) \subseteq (\lambda-\varepsilon, \lambda+\varepsilon)$. Since $\lim_{n\to\infty} \|f(H_n)\| = \|f(H)\| \geq 1$, it follows that $\mathrm{Spec}(H_n) \cap (\lambda - \varepsilon, \lambda + \varepsilon) \neq \emptyset$ for large enough n. □

Exercises

2.1 Prove that the set $\mathscr{A}$ defined in (2.1.2) is an algebra under pointwise multiplication.

2.2 Prove that $\langle x + y\rangle \leq 2\langle x\rangle\langle y\rangle$ for all $x, y \in \mathbf{R}^N$.

2.3 Prove that $|(\partial/\partial x_i)\langle x\rangle^\lambda| \leq |\lambda|\langle x\rangle^{\lambda-1}$ for all $x \in \mathbf{R}^N$ and $\lambda \in \mathbf{R}$. Use this to prove the bound

$$|\sigma_x(z) + i\sigma_y(z)| \leq c\langle x\rangle^{-1}\chi_U(z)$$

used in Lemma 2.2.1.

2.4 Prove that the function r_w of Lemma 2.2.7 lies in $\mathscr{A}$.

2.5 Prove that the function g of Lemma 2.2.8 lies in $\mathscr{A}$, by computing and then estimating the successive derivatives of g.

2.6 Find an explicit inductive formula for the computation of the determinant of a tridiagonal $n \times n$ matrix.

2.7 Let A and $\{A_n\}_{n=1}^\infty$ be bounded operators on a Banach space $\mathscr{B}$. Prove that $\text{s-lim}_{n\to\infty} A_n = A$ if and only if there exists a constant $c < \infty$ such that $\|A_n\| \leq c$ for all n, and also $\lim_{n\to\infty} \|A_n\phi - A\phi\| = 0$ for all ϕ in some dense subset of $\mathscr{B}$.

2.8 Prove that if $\text{s-lim}_{n\to\infty} A_n = A$ and $\text{s-lim}_{n\to\infty} B_n = B$ then $\text{s-lim}_{n\to\infty} A_n B_n = AB$.

2.9 Give an example of a sequence of rank one operators A_n on a Hilbert space such that $\text{s-lim}_{n\to\infty} A_n = 0$ but for which $\text{s-lim}_{n\to\infty} A_n^*$ does not exist.

2.10 Prove that if H, H_n are all non-negative self-adjoint operators on a Hilbert space $\mathscr{H}$, then H_n converges to H in the norm resolvent sense if and only if

$$\lim_{n\to\infty} \|(H_n + 1)^{-1} - (H + 1)^{-1}\| = 0.$$

2.11 Let $\{H_n\}_{n=1}^\infty$ be a sequence of self-adjoint operators which converge to H in the norm resolvent sense. If H has an isolated eigenvalue λ of multiplicity one, prove that for large enough n, H_n also have isolated eigenvalues λ_n of multiplicity one which converge to λ.

2.12 Find a counterexample to the statement of Theorem 2.6.2 if we assume not that $f \in C_0(\mathbf{R})$ but that f is the characteristic function of a bounded interval $[a, b]$.

2.13 Prove that if H is a self-adjoint operator on $\mathcal{H}$ and A is a bounded self-adjoint operator, then

$$(H + \lambda A - i)^{-1} = (H - i)^{-1} - (H + \lambda A - i)^{-1} \lambda A (H - i)^{-1}$$

for all $\lambda \in \mathbf{R}$. Deduce that $(H + \lambda A)$ converges in the norm resolvent sense to H as $\lambda \to 0$. Note: See Section 1.4 for the basic properties of the operator $(H + \lambda A)$.

2.14 Let H be a self-adjoint operator on a Hilbert space $\mathcal{H}$. Given $t \in \mathbf{R}$ we define e^{-iHt} to be the operator $f_t(H)$ where $f_t(s) := e^{-ist}$ for all $s \in \mathbf{R}$.

(a) Prove that each operator e^{-iHt} is unitary.

(b) Prove that $t \to e^{-iHt}$ is a unitary group, in the sense that

$$e^{-iH(t+s)} = e^{-iHt} e^{-iHs}$$

for all $s, t \in \mathbf{R}$.

(c) Prove that $t \to e^{-iHt}$ is strongly continuous, that is

$$\lim_{t \to a} \| e^{-iHt} f - e^{-iHa} f \| = 0$$

for all $a \in \mathbf{R}$ and all $f \in \mathcal{H}$.

(d) Prove that if $f \in \mathrm{Dom}(H)$ then $f(t) := e^{-iHt} f$ is a norm differentiable function of t and is a solution of the Schrödinger equation

$$\frac{\mathrm{d}f(t)}{\mathrm{d}t} = -iHf(t).$$

3
Translation invariant operators

3.1 Introduction

In this chapter we start the study of partial differential operators. We shall obtain a detailed spectral analysis of constant coefficient differential operators acting on the space $L^2 := L^2(\mathbf{R}^N, \mathrm{d}^N x)$. These operators are of interest both for their own sake and because the results can be used in the study of other partial differential operators. The theory is particularly complete because of the possibility of using the Fourier transform to obtain an explicit version of the spectral theorem 2.5.1.

We start by summarising the properties of the Fourier transform which we shall need. Our account uses the properties of Schwartz space, and the theory of distributions. However, we give a limited account of the theory of distributions, in which we do not make any assumption of continuity. This has the advantage of allowing us to avoid getting involved with any of the topological aspects of distribution theory. Since we shall not need any structure theorems for distributions, there is no disadvantage in our approach.

3.2 Schwartz space

Let $C^\infty := C^\infty(\mathbf{R}^N)$ denote the space of smooth functions on $\mathbf{R}^N$. Let α denote a multi-index $\{\alpha_1, ..., \alpha_N\}$ of non-negative integers and define the linear operator D^α on C^∞ by

$$D^\alpha f(x) := \frac{\partial^{\alpha_1}}{\partial x_1^{\alpha_1}} \cdots \frac{\partial^{\alpha_N}}{\partial x_N^{\alpha_N}} f(x_1, ..., x_N).$$

We say that the degree $|\alpha|$ of α is $|\alpha| = \alpha_1 + ... + \alpha_N$.

Given $x \in \mathbf{R}^N$ let $\langle x \rangle$ denote the quantity

$$\langle x \rangle := (1 + |x|^2)^{1/2}.$$

We define the Schwartz space $\mathscr{S}$ on $\mathbf{R}^N$ to be the set of $f \in \mathbf{C}^\infty$ such that

$$|D^\alpha f(x)| \le c_{\alpha,\beta}\langle x\rangle^{-\beta}$$

for all $x \in \mathbf{R}^N$, some $c_{\alpha,\beta} < \infty$, all multi-indices α and all $\beta > 0$. It is elementary that $\mathscr{S}$ is invariant under the action of D^γ for all multi-indices γ.

Clearly $\mathscr{S}$ contains the set $C_c^\infty := C_c^\infty(\mathbf{R}^N)$ of smooth functions on $\mathbf{R}^N$ which vanish if $|x|$ is sufficiently large. For many applications to partial differential equations C_c^∞ is more useful than $\mathscr{S}$. However, we shall see that $\mathscr{S}$ is very well behaved in connection with the Fourier transform, and this simplifies the considerations of this chapter.

Many computations involving smooth functions, such as the proof of our next lemma, are simplified by the use of a mollifier. This is a function k_s on $\mathbf{R}^N$ defined for each $s > 0$ as follows. We first let k be a smooth function on $\mathbf{R}^N$ such that

(1) $k(x) = 0$ if $|x| \ge 1$.
(2) $k(x) > 0$ if $|x| < 1$.
(3) $\int_{\mathbf{R}^N} k(x)\mathrm{d}^N x = 1$.

An example of such a function is

$$k(x) := \begin{cases} c\exp[-1/(1-|x|^2)] & \text{if } |x| < 1, \\ 0 & \text{otherwise,} \end{cases}$$

where $c > 0$ is chosen to satisfy (3). We then define k_s by

$$k_s(x) := s^{-N} k(x/s).$$

If $f, g \in L^1 := L^1(\mathbf{R}^N, \mathrm{d}^N x)$ then an application of Fubini's theorem shows that their convolution $f * g$, defined by

$$f * g(x) := \int_{\mathbf{R}^N} f(x-y)g(y)\mathrm{d}^N y,$$

is finite almost everywhere and again lies in L^1 with $\|f * g\|_1 \le \|f\|_1 \|g\|_1$. The next two lemmas show that a given function f can often be approximated arbitrarily closely by functions $k_s * f$ which are locally much more regular; see also Exercise 3.3.

Lemma 3.2.1 *If $f \in L^1$, then $\lim_{s\to 0} \|f - k_s * f\|_1 = 0$. Both $\mathscr{S}$ and C_c^∞ are norm dense in L^1.*

Proof If we define the linear operator T_s on L^1 by $T_s f := k_s * f$ then by estimating the relevant integrals one sees that

$$\|T_s g\|_1 \le \|k_s\|_1 \|g\|_1 = \|g\|_1$$

for all $g \in L^1$. Hence $\|T_s\| \le 1$ for all $s > 0$. The first statement therefore follows if we prove that $\lim_{s \to 0} \|T_s f - f\|_1 = 0$ for all f in a dense linear subspace of L^1. By linearity it suffices to prove this for the characteristic functions defined below.

If $a, b \in \mathbf{R}^N$ then we define $\chi_{[a,b]}$ to be the characteristic function of the 'interval'

$$[a, b] := \{x \in \mathbf{R}^N : a_r \le x_r \le b_r \text{ for all } 1 \le r \le N\}.$$

We assume that the reader is sufficiently familiar with Lebesgue integration to know that the set of all finite linear combinations of such characteristic functions is norm dense in L^1. We prove that every such characteristic function can be approximated arbitrarily closely in norm by functions in the smaller of the two spaces, namely C_c^∞.

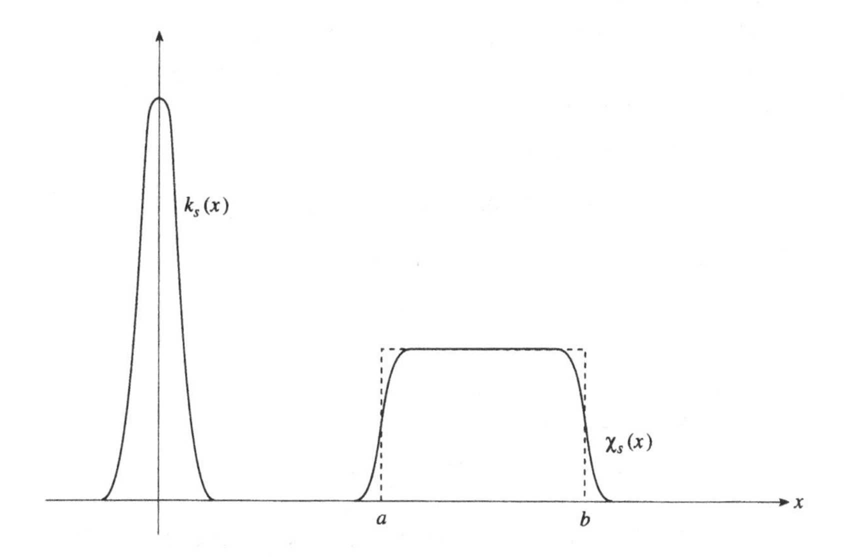

Putting $\chi := \chi_{[a,b]}$ and $\chi_s := \chi * k_s$, routine calculations establish the following facts:

(1) $0 \le \chi_s(x) \le 1$ for all $s > 0$ and all $x \in \mathbf{R}^N$.

(2) One has $\chi_s(x) = 0$ unless $\mathrm{dist}(x, [a, b]) < s$.

(3) If $x \notin [a, b]$ then $\lim_{s\to 0} \chi_s(x) = 0$.

(4) If x lies in the interior of $[a, b]$ then $\lim_{s\to 0} \chi_s(x) = 1$.

(5) Hence, using the dominated convergence theorem, χ_s converges to χ in L^1 norm as $s \to 0$.

(6) By differentiation under the integral sign one sees that χ_s is smooth. Thus it lies in C_c^∞. □

Lemma 3.2.2 *If f lies in the space L_c^2 of L^2 functions of compact support, then $k_s * f \in C_c^\infty$ for all $s > 0$ and*

$$\lim_{s\to 0} \|f - k_s * f\|_2 = 0. \tag{3.2.1}$$

Proof The proof that $k_s * f \in C_c^\infty$ depends upon verifying that one may differentiate under the integral sign. The proof that (3.2.1) holds if f is a step function on $\mathbf{R}^N$ follows the proof of Lemma 3.2.1 closely. If $g \in L^2$ then Exercise 3.14 states that

$$\|k_s * g\|_2 \le \|k_s\|_1 \|g\|_2 = \|g\|_2.$$

Therefore if f_n is a sequence of step functions converging to f we have

$$\begin{aligned}\limsup_{s\to 0} \|f - k_s * f\|_2 &\le \limsup_{s\to 0} \|f_n - k_s * f_n\|_2 \\ &\quad + \limsup_{s\to 0} \|f - f_n - k_s * f + k_s * f_n\|_2 \\ &\le 2\|f - f_n\|_2.\end{aligned}$$

Letting $n \to \infty$ we deduce that

$$\limsup_{s\to 0} \|f - k_s * f\|_2 = 0.$$ □

Given a multi-index α we define the monomial x^α of degree $|\alpha|$ to be the function $x_1^{\alpha_1} \ldots x_N^{\alpha_N}$ and define a polynomial of degree m to be a linear combination of monomials of the form

$$p(x) := \sum_{|\alpha| \le m} a_\alpha x^\alpha.$$

An application of Leibnitz' formula implies that $\mathcal{S}$ is invariant under multiplication by polynomials. Even more is actually true. Let $\mathscr{P}$ denote the space of all smooth functions f on $\mathbf{R}^N$ such that

$$|D^\alpha f(x)| \le c_\alpha \langle x \rangle^\beta$$

for some $\beta \ge 0$, all α and all $x \in \mathbf{R}^N$.

Lemma 3.2.3 *We have $\mathscr{S} \subseteq \mathscr{P}$. If $f \in \mathscr{S}$ and $g \in \mathscr{P}$, then $fg \in \mathscr{S}$.*

We omit the proof, which is elementary. We shall commonly not distinguish between a function $g \in \mathscr{P}$ and the multiplication operator on $\mathscr{S}$ associated with it.

Lemma 3.2.4 *If $f, g \in \mathscr{S}$ then $f * g \in \mathscr{S}$.*

One may give a direct computational proof of this lemma. A more sophisticated proof is provided in the next section.

3.3 The Fourier transform

The Fourier transform is easiest to define on the space L^1. In this case the integral

$$\hat{f}(y) := (2\pi)^{-N/2} \int_{\mathbf{R}^N} f(x) e^{-ix \cdot y} d^N x \tag{3.3.1}$$

is absolutely convergent for all $y \in \mathbf{R}^N$.

Lemma 3.3.1 *If $f \in L^1$, then $\hat{f}$ lies in the space $C_0(\mathbf{R}^N)$ of bounded continuous functions on $\mathbf{R}^N$ which vanish as $|y| \to \infty$.*

Proof It is immediate from its definition that

$$|\hat{f}(y)| \le (2\pi)^{-N/2} \int_{\mathbf{R}^N} |f(x)| d^N x$$

for all $y \in \mathbf{R}^N$; thus $\hat{f}$ is bounded. The continuity of $\hat{f}$ follows from the dominated convergence theorem. Finally the Riemann–Lebesgue lemma states that $\hat{f}(y) \to 0$ as $|y| \to \infty$. (The R–L lemma is proved by an explicit calculation if f is a finite linear combination of characteristic functions of intervals, and follows for general $f \in L^1$ by a density argument.) $\square$

One cannot define the Fourier transform so easily on L^2. The key result is the following.

Theorem 3.3.2 *The Fourier transform maps $\mathcal{S}$ into $\mathcal{S}$. If α is any multi-index and $f \in \mathcal{S}$, then*

$$(D^\alpha f)\hat{}(y) = (iy)^\alpha \hat{f}(y). \tag{3.3.2}$$

If we put $g(x) := x^\alpha f(x)$, then $g \in \mathcal{S}$ and

$$\hat{g}(y) = i^{|\alpha|} D^\alpha \hat{f}(y) \tag{3.3.3}$$

for all $y \in \mathbf{R}^N$.

Proof If $\hat{\mathcal{S}}$ denotes $\{\hat{f} : f \in \mathcal{S}\}$, then $\hat{\mathcal{S}} \subseteq C_0(\mathbf{R}^N)$ by Lemma 3.3.1.

If we integrate (3.3.1) by parts with respect to the variable x_r, then after checking that the limits as $x_r \to \pm\infty$ vanish we obtain

$$\begin{aligned}
-iy_r \hat{f}(y) &= (2\pi)^{-N/2} \int_{\mathbf{R}^N} f(x) \frac{\partial \mathrm{e}^{-ix\cdot y}}{\partial x_r} \mathrm{d}^N x \\
&= -(2\pi)^{-N/2} \int_{\mathbf{R}^N} \frac{\partial f}{\partial x_r} \mathrm{e}^{-ix\cdot y} \mathrm{d}^N x \\
&= -(D_r f)\hat{}(y)
\end{aligned}$$

for all $y \in \mathbf{R}^N$, where $D_r := \partial/\partial x_r$. Multiple applications of this calculation yield both (3.3.2) and the fact that $y^\alpha \hat{f} \in \hat{\mathcal{S}}$ for all $f \in \mathcal{S}$.

The proof of (3.3.3) depends upon differentiating (3.3.1) under the integral sign repeatedly – the conditions permitting this are easy to verify. We obtain

$$\begin{aligned}
D^\alpha \hat{f}(y) &= (2\pi)^{-N/2} \int_{\mathbf{R}^N} f(x) D^\alpha_{(y)} \mathrm{e}^{-ix\cdot y} \mathrm{d}^N x \\
&= (2\pi)^{-N/2} \int_{\mathbf{R}^N} f(x)(-i)^{|\alpha|} x^\alpha \mathrm{e}^{-ix\cdot y} \mathrm{d}^N x \\
&= (-i)^{|\alpha|} \hat{g}(y)
\end{aligned}$$

for all $y \in \mathbf{R}^N$. This is equivalent to (3.3.3). This calculation also implies that $D^\alpha \hat{f} \in \hat{\mathcal{S}}$ for all $f \in \mathcal{S}$.

If α and β are any two multi-indices, then we conclude that $y^\beta D^\alpha \hat{f}(y) \in \hat{\mathcal{S}}$ for all $f \in \mathcal{S}$. Therefore there exists a constant $c_{\alpha,\beta} < \infty$ such that

$$|y^\beta D^\alpha \hat{f}(y)| \le c_{\alpha,\beta}$$

for all $y \in \mathbf{R}^N$. This is enough to establish that $\hat{\mathcal{S}} \subseteq \mathcal{S}$; see Exercise 3.2. □

Theorem 3.3.3 *The Fourier transform $f \to \hat{f}$ maps $\mathcal{S}$ one-one onto $\mathcal{S}$. It preserves the L^2 norm and L^2 inner product, and extends uniquely to* *a unitary map $\mathcal{F}$ of L^2 onto L^2. The inverse map is the unique unitary extension to L^2 of the map $f \to \check{f}$ defined on $\mathcal{S}$ by*

$$\check{f}(x) := (2\pi)^{-N/2} \int_{\mathbf{R}^N} f(y) e^{ix \cdot y} \mathrm{d}^N y.$$

Proof The method of separation of variables establishes that for all $t > 0$ the function

$$k_t(x) := (4\pi t)^{-N/2} \exp[-|x|^2/4t] \tag{3.3.4}$$

satisfies

$$\int_{\mathbf{R}^N} \exp[-ix \cdot y - t|y|^2] \mathrm{d}^N y = (2\pi)^N k_t(x). \tag{3.3.5}$$

The function k_t is not quite a mollifier in the sense of the last section, but plays a similar role. If f is a bounded continuous function on $\mathbf{R}^N$ such that both f and $\hat{f}$ lie in L^1, then an application of Fubini's theorem establishes that

$$\begin{aligned}
&\int_{\mathbf{R}^N} \hat{f}(y) \exp[-t|y|^2] \mathrm{d}^N y \\
&\quad = (2\pi)^{-N/2} \int_{\mathbf{R}^N} \int_{\mathbf{R}^N} f(x) \exp[-ix \cdot y - t|y|^2] \mathrm{d}^N x \mathrm{d}^N y \\
&\quad = (2\pi)^{N/2} \int_{\mathbf{R}^N} f(x) k_t(x) \mathrm{d}^N x \\
&\quad = (2\pi)^{N/2} \int_{\mathbf{R}^N} f(t^{1/2} x) k_1(x) \mathrm{d}^N x.
\end{aligned}$$

Letting $t \to 0$, the dominated convergence theorem now yields

$$\begin{aligned}
\int_{\mathbf{R}^N} \hat{f}(y) \mathrm{d}^N y &= (2\pi)^{N/2} \int_{\mathbf{R}^N} f(0) k_1(x) \mathrm{d}^N x. \\
&= (2\pi)^{N/2} f(0).
\end{aligned} \tag{3.3.6}$$

Replacing the function $x \to f(x)$ by $x \to f(x+u)$ we obtain

$$f(u) = (2\pi)^{-N/2} \int_{\mathbf{R}^N} \hat{f}(y) \exp[iu \cdot y] \mathrm{d}^N y \tag{3.3.7}$$

for all $u \in \mathbf{R}^N$.

If $f \in \mathcal{S}$, then Theorem 3.3.2 states that $g := \hat{f} \in \mathcal{S} \subseteq L^1$ and the above argument then yields $\check{g} = f$. It follows that $\hat{}$ maps $\mathcal{S}$ one-one onto $\mathcal{S}$ and that $\check{}$ is the inverse map.

If $f, g \in \mathcal{S}$ then a direct computation establishes that

$$(f * g)\hat{}(y) = (2\pi)^{N/2}\hat{f}(y)\hat{g}(y). \tag{3.3.8}$$

By applying (3.3.6) to $f * g$ we now obtain

$$\begin{aligned}\int_{\mathbf{R}^N} f(x)g(-x)\mathrm{d}^N x &= (f * g)(0) \\ &= (2\pi)^{-N/2}\int_{\mathbf{R}^N}(f * g)\hat{}(y)\mathrm{d}^N y \\ &= \int_{\mathbf{R}^N}\hat{f}(y)\hat{g}(y)\mathrm{d}^N y.\end{aligned}$$

Replacing $x \to g(x)$ by $x \to \overline{g(-x)}$ we obtain

$$\int_{\mathbf{R}^N} f(x)\overline{g(x)}\mathrm{d}^N x = \int_{\mathbf{R}^N}\hat{f}(y)\overline{\hat{g}(y)}\mathrm{d}^N y$$

for all $f, g \in \mathcal{S}$. This establishes that $\hat{}$ preserves the L^2 norm and L^2 inner product. It may be extended in a unique manner to a unitary operator $\mathcal{F}$ on L^2 since $\mathcal{S}$ is norm dense in L^2 by Lemma 3.2.2. □

Proof of Lemma 3.2.4 Given $f, g \in \mathcal{S}$, an application of (3.3.8) establishes that $(f * g)\hat{} \in \mathcal{S}$. Theorem 3.3.3 now implies that $f * g \in \mathcal{S}$. □

It is worth commenting that the above proof is also the basis of an efficient numerical method of computing convolutions. In 1965 Cooley and Tukey devised a very fast method of computing Fourier transforms, called the FFT for obvious reasons, which has transformed this part of numerical analysis. Instead of computing the convolution of two functions directly, it is actually faster to evaluate the Fourier transform of each function, and then take the inverse transform of their pointwise product, including the appropriate power of 2π. The FFT may also be used to solve convolution equations, which arise in the important problem of image enhancement.

3.4 Distributions

Several mathematicians, most notably Sobolev in 1936, contributed to the study of weak solutions of differential equations. Without doubt, however, the credit for putting these ideas together in a coherent theory of distributions should go to Schwartz (1950). His theory laid the framework for much subsequent analysis of a more detailed character, by shifting the emphasis from the proof of the existence of solutions of various equations to the proof of regularity properties of weak solutions.

In this section we give a brief account of Schwartz' theory of distributions on $\mathbf{R}^N$. Our version of this theory is unorthodox in that we do not impose any continuity conditions upon distributions. However, we do not go far enough with the theory for this to be an embarrassment, and all of our proofs are the standard ones, but with the topology removed. The only reason for our introducing the ideas is to enable us to add together functions from different L^p spaces of functions and take Fourier transforms of the results without worrying about the correctness and consistency of this procedure.

We define a distribution to be any linear functional ϕ from $\mathscr{S}$ to $\mathbf{C}$. If $f \in \mathscr{S}$ we use the notation $\phi(f)$ interchangably with $\langle \phi, f \rangle$; in contrast with the situation for inner products, this expression is complex linear in both variables. We regard L^p for $1 \leq p \leq \infty$ as a linear subspace of the space $\mathscr{S}'$ of distributions by identifying $g \in L^p$ with the linear functional

$$\phi_g(f) := \int_{\mathbf{R}^N} f(x)g(x)\mathrm{d}^N x. \tag{3.4.1}$$

The 'Dirac delta function' at 0 is the distribution $f \to f(0)$ where $f \in \mathscr{S}$. This terminology is an example of the widespread and harmless practice of referring to distributions as if they were 'generalised' functions on $\mathbf{R}^N$ rather than linear functionals on $\mathscr{S}$.

Lemma 3.4.1 *The formula*

$$\hat{\phi}(f) := \phi(\hat{f})$$

defines a one-one linear map of $\mathscr{S}'$ onto $\mathscr{S}'$ which is consistent with the definition of the Fourier transform on L^1 and L^2. In particular if $f \in L^1 \cap L^2$, then the L^1 and L^2 definitions of $\hat{f}$ are consistent with each other.

Proof Since $\hat{}$ defines a one-one map of $\mathscr{S}$ onto itself, the above formula defines a similar map on $\mathscr{S}'$. Let ϕ_g be the distribution associated with a function $g \in L^1$. Then

$$\begin{aligned}(\phi_g)\hat{}(f) &= \phi_g(\hat{f}) \\ &= \int_{\mathbf{R}^N} \hat{f}(x) g(x) \mathrm{d}^N x \\ &= (2\pi)^{-N/2} \int_{\mathbf{R}^N} \int_{\mathbf{R}^N} f(y) \mathrm{e}^{-ix\cdot y} g(x) \mathrm{d}^N x \mathrm{d}^N y \\ &= \int_{\mathbf{R}^N} f(x) \hat{g}(x) \mathrm{d}^N x \\ &= \phi_{\hat{g}}(f)\end{aligned}$$

for all $f \in \mathcal{S}$. This establishes that $(\phi_g)\hat{} = \phi_{\hat{g}}$ for all $g \in L^1$.

If $g \in L^2$, then there exists a sequence $g_n \in L^1 \cap L^2$ such that $\|g_n - g\|_2 \to 0$ as $n \to \infty$. Letting $n \to \infty$ in the identity

$$\phi_{\hat{g}_n}(f) = \phi_{g_n}(\hat{f}),$$

valid for all $f \in \mathcal{S}$, yields $(\phi_g)\hat{} = \phi_{\hat{g}}$ for all $g \in L^2$. □

We now define the differential operator D^α on $\mathcal{S}'$ and the multiplication operator associated with $g \in \mathcal{P}$ on $\mathcal{S}'$ by

$$\langle D^\alpha \phi, f \rangle := (-1)^{|\alpha|} \langle \phi, D^\alpha f \rangle \quad , \quad \langle g\phi, f \rangle := \langle \phi, gf \rangle.$$

If $f \in L^p$ for some $1 \le p \le \infty$, then its weak derivative $D^\alpha f$ refers to the distribution obtained by differentiating f in the above sense; this always exists whether or not f is differentiable in the classical sense. The similarity between our next lemma and Theorem 3.3.2 provides justification for regarding the operations just defined as extensions of the 'same' operations on $\mathcal{S}$.

Lemma 3.4.2 *If $\phi \in \mathcal{S}'$, then*

$$\begin{aligned}(D^\alpha \phi)\hat{} &= i^{|\alpha|} x^\alpha \hat{\phi}, \\ (x^\alpha \phi)\hat{} &= i^{|\alpha|} D^\alpha \hat{\phi}.\end{aligned}$$

Proof If $f \in \mathcal{S}$ and $\phi \in \mathcal{S}'$, then

$$\begin{aligned}\langle (D^\alpha\phi)\hat{}, f\rangle &= \langle D^\alpha\phi, \hat{f}\rangle \\ &= (-1)^{|\alpha|}\langle \phi, D^\alpha\hat{f}\rangle \\ &= i^{|\alpha|}\langle \phi, (x^\alpha f)\hat{}\rangle \\ &= i^{|\alpha|}\langle \hat{\phi}, x^\alpha f\rangle \\ &= i^{|\alpha|}\langle x^\alpha\hat{\phi}, f\rangle.\end{aligned}$$

This identity establishes the first statement of the lemma. Also

$$\begin{aligned}\langle (x^\alpha\phi)\hat{}, f\rangle &= \langle \phi, x^\alpha\hat{f}\rangle \\ &= \langle \phi, (-i)^{|\alpha|}(D^\alpha f)\hat{}\rangle \\ &= \langle \hat{\phi}, (-i)^{|\alpha|}D^\alpha f\rangle \\ &= i^{|\alpha|}\langle D^\alpha\hat{\phi}, f\rangle.\end{aligned}$$

This identity establishes the second statement of the lemma. □

If ϕ is a distribution on $\mathbf{R}^N$ and g is a function which is locally integrable on an open subset U of $\mathbf{R}^N$, then we say that $\phi = g$ on U if

$$\phi(f) = \int_U f(x)g(x)\mathrm{d}^N x$$

for all $f \in C_c^\infty$ which have support inside U. Thus the Dirac delta function equals 0 on the complement of the origin, but so also does the distribution δ_α defined for any multi-index α by

$$\delta_\alpha(f) := (D^\alpha f)(0).$$

In Chapter 2 we defined S^β for $\beta \in \mathbf{R}$ to be a certain space of functions on $\mathbf{R}$. In N dimensions we define it to be the space of smooth functions f on $\mathbf{R}^N$ such that for every multi-index α there exists $c_\alpha < \infty$ for which

$$|D^\alpha f(x)| \le c_\alpha\langle x\rangle^{\beta-|\alpha|} < \infty$$

for all $x \in \mathbf{R}^N$.

Theorem 3.4.3 *If $f \in S^\beta$ for some $\beta < 0$ then its Fourier transform $\hat{f}$ in the sense of distributions is a continuous function on the complement of the origin which decreases rapidly at infinity in the sense that for all $\gamma > 0$ there exists $c_\gamma < \infty$ such that*

$$|\hat{f}(y)| \le c_\gamma|y|^{-\gamma}$$

for all $|y| \ge 1$.

Proof Let f denote both the function in S^β and the associated distribution. If $|\alpha| \geq N$ then $D^\alpha f \in L^1$. Taking Fourier transforms and using Theorem 3.4.2 we deduce that $y^\alpha \hat{f}(y) \in C_0(\mathbf{R}^N)$. Therefore there exist constants $\tilde{c}_{|\alpha|} < \infty$ such that

$$|y_r|^{|\alpha|} |\hat{f}(y)| \leq \tilde{c}_{|\alpha|}$$

for all $y \in \mathbf{R}^N$, all $1 \leq r \leq N$ and all α such that $|\alpha| \geq N$. This implies the statement of the theorem. □

The behaviour of the Fourier transform near the origin is most easily controlled under another hypothesis.

Theorem 3.4.4 *If f is continuous and there exist constants $c_r \in \mathbf{C}$ and $a_r > 0$ for $1 \leq r \leq m$, and a constant $\varepsilon > 0$ such that*

$$f(x) - \sum_{r=1}^{m} c_r \langle x \rangle^{-a_r} = O(|x|^{-N-\varepsilon}) \tag{3.4.2}$$

as $|x| \to \infty$, then the Fourier transform of f lies in $L^1 + L^\infty$.

Proof Under the hypotheses of the theorem the left-hand side of (3.4.2) is a continuous function in L^1, so its Fourier transform lies in $C_0(\mathbf{R}^N) \subseteq L^\infty$. We shall complete the proof by showing that for any $a > 0$ the function $x \to \langle x \rangle^{-a}$ has a Fourier transform in L^1.

We define $g_a(y) > 0$ for $y \in \mathbf{R}^N$ by

$$g_a(y) := \Gamma(a)^{-1} \int_{t=0}^{\infty} k_t(y) \mathrm{e}^{-t} t^{a-1} \mathrm{d}t$$

where $k_t(y)$ is defined in (3.3.4). It is elementary that

$$\int_{\mathbf{R}^N} g_a(y) \mathrm{d}^N y = \Gamma(a)^{-1} \int_{t=0}^{\infty} \mathrm{e}^{-t} t^{a-1} \mathrm{d}t = 1,$$

so $g_a \in L^1$. If $x \in \mathbf{R}^N$, we also have

$$\begin{aligned}
\int_{\mathbf{R}^N} g_a(y) \exp[ix \cdot y] \mathrm{d}^N y &= \Gamma(a)^{-1} \int_{\mathbf{R}^N} \int_{t=0}^{\infty} k_t(y) \exp[ix \cdot y - t] t^{a-1} \mathrm{d}t \, \mathrm{d}^N y \\
&= \Gamma(a)^{-1} \int_{t=0}^{\infty} \exp[-t|x|^2 - t] t^{a-1} \mathrm{d}t \\
&= (|x|^2 + 1)^{-a}.
\end{aligned}$$

Therefore $x \to \langle x \rangle^{-a}$ is the inverse Fourier transform of $(2\pi)^{N/2} g_a(y)$, which lies in L^1. The result follows by the Fourier inversion theorem. □

Corollary 3.4.5 *Let*

$$f(x) := \sum_{r=1}^{m} c_r \langle x \rangle^{-a_r} + g(x)$$

for all $x \in \mathbf{R}^N$, *where* $c_r \in \mathbf{C}$, $a_r > 0$ *and* $g \in S^\beta$ *for some* $\beta < -N$. *Then* $\hat{f} \in L^1$.

Proof By Theorem 3.4.3 we see that

$$\int_{|y|>1} |\hat{f}(y)| \mathrm{d}^N y < \infty$$

and by Theorem 3.4.4 we see that

$$\int_{|y|<1} |\hat{f}(y)| \mathrm{d}^N y < \infty. \qquad \square$$

3.5 Differential operators

A linear differential operator on $\mathbf{R}^N$ is defined to be an operator of the form

$$f \to Lf := \sum_\alpha a_\alpha(x) D^\alpha f(x). \tag{3.5.1}$$

We assume that the sum is finite and that the domain of L is $\mathscr{S}$; all of the results in this section can be adapted to the case where L has domain C_c^∞ by the application of Theorem 3.7.3. The maximum value of $|\alpha|$ involved in the above sum is called the order of the operator. We say that such an operator is translation invariant if $T_u L = L T_u$ for all $u \in \mathbf{R}^N$, where

$$T_u f(x) := f(x+u).$$

Lemma 3.5.1 *If the coefficients* $a_\alpha(x)$ *lie in* $\mathscr{P}$ *then the differential operator* L *defined by (3.5.1) maps* $\mathscr{S}$ *into* $\mathscr{S}$. *This operator is translation invariant if and only if the coefficients are constant.*

Proof The first statement of the lemma follows immediately from Lemma 3.2.3. If L is translation invariant, then

$$\begin{aligned} 0 &= T_u L T_{-u} f - Lf \\ &= \sum_\alpha \{a_\alpha(x+u) - a_\alpha(x)\} D^\alpha f(x) \end{aligned}$$

for all f in the domain of L. Now fix α, and choose $f \in C_c^\infty$ such that $D^\alpha f(0) = 1$ but $D^\beta f(0) = 0$ for all $\beta \neq \alpha$ which occur in the sum; we leave the construction of such an f as an exercise. For this choice of f the above equation yields $0 = a_\alpha(u) - a_\alpha(0)$ for all $u \in \mathbf{R}^N$, so a_α is constant. The converse is elementary. □

The symbol of a differential operator of the form (3.5.1) is defined to be the function

$$a(x, y) := \sum_\alpha a_\alpha(x)(iy)^\alpha \tag{3.5.2}$$

of the two variables x, y in $\mathbf{R}^N$. For a constant coefficient operator the symbol is the polynomial

$$a(y) := \sum_\alpha a_\alpha(iy)^\alpha. \tag{3.5.3}$$

There is a close connection between symbols, Fourier transforms and the ideas about multiplication operators described in Section 1.3.

Lemma 3.5.2 *If L has symbol $a(x, y)$ and $f \in \mathcal{S}$, then*

$$Lf(x) = (2\pi)^{-N/2} \int_{\mathbf{R}^N} a(x, y)\hat{f}(y)e^{ix\cdot y}\mathrm{d}^N y$$

for all $x \in \mathbf{R}^N$.

Proof It follows from Theorems 3.3.2 and 3.3.3 that

$$\begin{aligned}
Lf(x) &= \sum_\alpha a_\alpha(x) D^\alpha f(x) \\
&= \sum_\alpha a_\alpha(x)(2\pi)^{-N/2} \int_{\mathbf{R}^N} (D^\alpha f)\hat{}(y)e^{ix\cdot y}\mathrm{d}^N y \\
&= \sum_\alpha a_\alpha(x)(2\pi)^{-N/2} \int_{\mathbf{R}^N} (iy)^\alpha \hat{f}(y)e^{ix\cdot y}\mathrm{d}^N y \\
&= (2\pi)^{-N/2} \int_{\mathbf{R}^N} a(x, y)\hat{f}(y)e^{ix\cdot y}\mathrm{d}^N y.
\end{aligned}$$ □

The following theorem is also valid if we replace $\mathcal{S}$ by C_c^∞. The proof involves combining the theorem as stated with Theorem 3.7.3.

Theorem 3.5.3 *The constant coefficient differential operator H with domain $\mathcal{S}$ is symmetric if and only if its symbol is real-valued on $\mathbf{R}^N$. In this case* *the closure $\overline{H}$ of the operator is self-adjoint and its spectrum is the closure of the set*

$$\left\{ \sum_{\alpha} a_\alpha (iy)^\alpha : y \in \mathbf{R}^N \right\}.$$

Proof Putting $H = L$ in Lemma 3.5.2, we have

$$Lf(x) = (2\pi)^{-N/2} \int_{\mathbf{R}^N} a(y)\hat{f}(y)e^{ix \cdot y} d^N y$$

for all $f \in \mathcal{S}$. Since $\hat{}$ is a unitary operator from $L^2(\mathbf{R}^N, d^N x)$ to $L^2(\mathbf{R}^N, d^N x)$, this formula is precisely analogous to Theorem 2.5.1, where the corresponding unitary operator is U and the multiplication operator is h. We have only to prove all of the statements of the theorem for the operator $A := \mathcal{F} H \mathcal{F}^{-1}$ of multiplication by $a(y)$ acting on $\mathcal{S}$.

The fact that A is symmetric if and only if a is real-valued is evident. Assuming that this is true, we define the multiplication operator B on L^2 by $Bf := af$ with domain

$$\mathcal{D} := \left\{ f : \int_{\mathbf{R}^N} (1 + a(y)^2) |f(y)|^2 d^N y < \infty \right\}.$$

It follows from Lemma 1.3.1 that B is self-adjoint on its domain, while Lemma 1.3.2 allows us to identify its spectrum with the essential range of the function a.

All that remains is to prove that B is the closure of the operator A. By Lemma 1.3.1 B is the closure of its restriction to L^2_c. We therefore have only to show that for any $f \in L^2_c$ there exists a sequence $f_n \in \mathcal{S}$ such that $\|f - f_n\|_2 \to 0$ and $\|af - af_n\|_2 \to 0$ as $n \to \infty$. This is achieved by putting $f_n := f * k_{1/n}$, where k_s is the mollifier defined in Section 3.2, and using Lemma 3.2.2. □

If a differential operator L is of order n then its principal symbol is defined to be

$$a_n(x, y) := \sum_{|\alpha|=n} a_\alpha(x)(iy)^\alpha.$$

The operator L of even order n is said to be uniformly elliptic if there is a constant $c > 0$ such that

$$c^{-1}|y|^n \le a_n(x, y) \le c|y|^n$$

for all $x, y \in \mathbf{R}^N$.

Of particular interest is the case where

$$Lf := -\sum_{i,j=1}^{N} a_{i,j}(x)\frac{\partial^2 f}{\partial x_i \partial x_j} + \sum_{j=1}^{N} b_j(x)\frac{\partial f}{\partial x_j}$$

and where $a(x) := \{a_{i,j}(x)\}$ is a real symmetric matrix for every $x \in \mathbf{R}^N$. The principal symbol is then

$$a_2(x, y) := \sum_{i,j=1}^{N} a_{i,j}(x) y_i y_j.$$

The operator L is elliptic if all the eigenvalues of $a(x)$ are positive for all $x \in \mathbf{R}^N$, and is uniformly elliptic if there exists $c > 0$ such that

$$c^{-1}1 \leq a(x) \leq c1$$

in the sense of matrices, for all $x \in \mathbf{R}^N$; for constant coefficient operators there is, of course, no distinction between these notions.

Corollary 3.5.4 *If the constant coefficient differential operator H on $\mathcal{S}$ is symmetric and elliptic, then the spectrum of its closure is an interval of the form $[\lambda, +\infty)$.*

Proof We may write the symbol of H in the form

$$a(y) = a_n(y) + b(y),$$

where $b(y)$ is the sum of the terms of order less than the order n of H. Since $a_n(y) \geq c^{-1}|y|^n$ and $|b(y)| \leq c'(|y|^{n-1} + 1)$ for all $y \in \mathbf{R}^N$, it follows that $\lim_{|y|\to\infty} a(y) = +\infty$. Therefore the function $a(\cdot)$ is bounded below and its essential range is of the stated form. The proof is completed by applying Lemma 1.3.2. □

Theorem 3.5.5 *Let H be a constant coefficient symmetric differential operator with domain $\mathcal{S}$ and let f be a continuous bounded function on the real line. Then*

$$f(\overline{H})\phi = \mathcal{F}^{-1} M \mathcal{F} \phi$$

for all $\phi \in L^2(\mathbf{R}^N)$, where $\mathcal{F}$ denotes the Fourier transform operator and M is the multiplication operator given by

$$M\psi(y) := f\{\sum_{\alpha} a_\alpha (iy)^\alpha\}\psi(y)$$

for all $y \in \mathbf{R}^N$ and $\psi \in L^2(\mathbf{R}^N)$.

The proof is a direct application of Theorem 2.5.3. Whether or not one can write

$$f(\overline{H})\phi = k_f * \phi$$

depends upon whether there exists a function k_f on $\mathbf{R}^N$ whose Fourier transform is $f\{\sum_\alpha a_\alpha(iy)^\alpha\}$.

Example 3.5.6 There are a few cases in which one can find explicit functions g_λ such that $(\lambda^2 + H)^{-1}f = g_\lambda * f$ for all $f \in L^2$. Let $H_0 := -\overline{\Delta}$ acting upon $L^2(\mathbf{R}^N)$. The symbol of this operator is $a(y) = |y|^2$ and its spectrum is $[0, \infty)$. The operator commutes with rotations about the origin, and all operators $f(H_0)$ therefore also have this property. We claim that if $N = 3$ the functions g_λ are given by

$$g_\lambda(x) = \mathrm{e}^{-\lambda|x|}/(4\pi|x|).$$

The proof of this statement starts with the observation that g_λ is a rotationally invariant function in $L^1(\mathbf{R}^3)$. Therefore its Fourier transform is also rotationally invariant and bounded. Using spherical polar coordinates we have

$$\begin{aligned}
&\int_{\mathbf{R}^3} g_\lambda(x)\exp[-ix\cdot y]\mathrm{d}^3x \\
&\qquad = \iiint \frac{1}{4\pi r}\exp[-r\lambda - ir|y|\cos\phi]r^2\sin\phi\,\mathrm{d}r\,\mathrm{d}\theta\,\mathrm{d}\phi \\
&\qquad = \int_0^\infty \frac{\mathrm{e}^{-r\lambda}}{2r}\left[\frac{\mathrm{e}^{-ir|y|\cos\phi}}{ir|y|}\right]_0^\pi r^2\mathrm{d}r \\
&\qquad = \int_0^\infty \frac{\mathrm{e}^{-r\lambda}}{2i|y|}\left[\mathrm{e}^{ir|y|} - \mathrm{e}^{-ir|y|}\right]\mathrm{d}r \\
&\qquad = \frac{1}{\lambda^2 + |y|^2}.
\end{aligned}$$

Similar elementary expressions can be obtained for this problem in all odd dimensions, but they become rapidly more complicated as the dimension increases. In even dimensions the corresponding functions are Bessel functions. Without knowing the exact formulae it is possible on general principles to see that the functions are rotationally invariant. □

3.6 Some L^p estimates

In this section we collect together a number of technical estimates concerning the L^p spaces. Such estimates are of great importance, but it is easy to get so deeply involved in such matters as Sobolev embedding theorems, weak L^p spaces, interpolation theorems, etc., that one loses sight of the differential operators for which all the work is being done. Although we have deliberately reduced the weight of this material, we need some of the results, particularly in Chapter 8.

We assume that the reader is familar with the following facts. If $1 \leq p < \infty$, then the space $L^p := L^p(\mathbf{R}^N)$ of all measurable functions $f : \mathbf{R}^N \to \mathbf{C}$ with finite norms

$$\|f\|_p := \left\{ \int_{\mathbf{R}^N} |f(x)|^p \mathrm{d}^N x \right\}^{1/p}$$

forms a Banach space provided one identifies two functions which are equal almost everywhere. The space L^∞ is defined as the Banach space of measurable functions for which

$$\|f\|_\infty := \min\{c \geq 0 : \mathrm{meas}\{x : |f(x)| > c\} = 0\}$$

is finite, again identifying two functions if they are equal almost everywhere. If $1 \leq p < \infty$, $1/p + 1/q = 1$ and $f \in L^p$ then the formula

$$\phi(f) := \langle f, g \rangle = \int_{\mathbf{R}^N} f(x) g(x) \mathrm{d}^N x$$

provides a one-one norm-preserving map between $\phi \in (L^p)^*$ and $g \in L^q$. The space $(L^\infty)^*$ is not easy to describe and will not be used.

Lemma 3.6.1 *Let $f \in L^p$ and $g \in L^q$ where $1 \leq p \leq \infty$, $1 \leq q \leq \infty$ and assume that $1/p + 1/q = 1/s$ where $1 \leq s \leq \infty$. Then $fg \in L^s$ and*

$$\|fg\|_s \leq \|f\|_p \|g\|_q.$$

Proof We assume that $p < \infty$ and $q < \infty$, the exceptional cases being easy to deal with. If we define $F : [0,1] \to [0, \infty)$ by

$$F(t) := \int_{\mathbf{R}^N} |f(x)|^{pt} |g(x)|^{q(1-t)} \mathrm{d}^N x,$$

then $F(0) = \|g\|_q^q$ and $F(1) = \|f\|_p^p$.

By rewriting $F(t)$ in the form

$$F(t) = \int_{\mathbf{R}^N} \mathrm{e}^{a(x)t} \mu(\mathrm{d}x)$$

where $a(x)$ is a real-valued function and μ is a non-negative measure, it is straightforward to show that

$$\frac{\mathrm{d}^2}{\mathrm{d}t^2}\log F(t) = \frac{F F'' - (F')^2}{F^2}$$

for all $0 < t < 1$. Since the numerator is non-negative by the Schwarz inequality, one sees that $\log F$ is convex and continuous on $[0,1]$. Putting $t := q/(p+q)$ and $s := pq/(p+q)$ we obtain

$$\begin{aligned}\|fg\|_s^s &= F\Big(\frac{q}{p+q}\Big)\\ &\le F(0)^{p/(p+q)}F(1)^{q/(p+q)}\\ &= \|f\|_p^s\|g\|_q^s.\end{aligned}$$

□

In order to state our next theorem, we need some notation. Let $1 \le p_0, p_1 \le \infty$ and $1 \le q_0, q_1 \le \infty$. Given $0 < \lambda < 1$, define p and q by

$$\frac{1}{p} := \frac{1-\lambda}{p_0} + \frac{\lambda}{p_1}, \quad \frac{1}{q} := \frac{1-\lambda}{q_0} + \frac{\lambda}{q_1}.$$

Let $\mathscr{D}$ denote the space of all simple functions on $\mathbf{R}^N$, that is all functions of the form

$$f(x) = \sum_{r=1}^{n} \alpha_r \chi_{E_r}(x),$$

where $\alpha_r \in \mathbf{C}$ and χ_{E_r} are the characteristic functions of disjoint sets $E_r \subseteq \mathbf{R}^N$ of finite measure.

The following interpolation theorem is of fundamental importance in many areas of analysis.

Theorem 3.6.2 *Let $T : \mathscr{D} \to L^{q_0} \cap L^{q_1}$ be a linear operator satisfying*

$$\|Tf\|_{q_i} \le c_i \|f\|_{p_i}$$

for all $f \in \mathscr{D}$ and $i = 0, 1$. Then T can be extended to a bounded linear operator $\overline{T} : L^p \to L^q$ such that

$$\|\overline{T}\|_{L^p \to L^q} \le c_0^{1-\lambda} c_1^{\lambda}$$

for all $0 < \lambda < 1$.

Proof We define q^* by $1/q + 1/q^* = 1$ with similar definitions of q_0^* and q_1^*. We also assume that p_0, p_1, q_0^* and q_1^* are all finite, leaving the reader to treat the exceptional cases.

Let $f, g \in \mathscr{D}$ have expansions

$$f(x) := \sum_{r=1}^{m} \alpha_r e^{i\theta_r} \chi_{E_r}(x),$$

where $\alpha_r > 0$, $\theta_r \in \mathbf{R}$ and $E_r \subseteq \mathbf{R}^N$ are disjoint sets of finite measure, and

$$g(x) := \sum_{s=1}^{n} \beta_s e^{i\phi_s} \chi_{F_s}(x),$$

where $\beta_s > 0$, $\phi_s \in \mathbf{R}$ and $F_s \subseteq \mathbf{R}^N$ are disjoint sets of finite measure. The entire proof is based upon an analysis of the function $G(z)$ defined on $S := \{z : 0 \le \mathrm{Re}(z) \le 1\}$ by

$$G(z) := \sum_r \sum_s \alpha_r^{uz+v} \beta_s^{wz+t} e^{i\theta_r} e^{-i\phi_s} \langle T\chi_{E_r}, \chi_{F_s} \rangle,$$

where $v := p/p_0$, $u := p/p_1 - p/p_0$, $t := q^*/q_0^*$ and $w := q^*/q_1^* - q^*/q_0^*$. We shall apply Hadamard's three lines lemma (from analytic function theory) to G, noting that G is bounded and continuous on S and analytic in the interior of S.

We first observe that if $y \in \mathbf{R}$, then

$$\begin{aligned}
|G(iy)| &= \left| \left\langle T\Big(\sum_r \alpha_r^{iuy+v} e^{i\theta_r} \chi_{E_r}\Big), \Big(\sum_s \beta_s^{iwy+t} e^{i\phi_s} \chi_{F_s}\Big) \right\rangle \right| \\
&\le c_0 \Big\| \sum_r \alpha_r^v \chi_{E_r} \Big\|_{p_0} \Big\| \sum_s \beta_s^t \chi_{F_s} \Big\|_{q_0^*} \\
&= c_0 \|f\|_{vp_0}^v \|g\|_{tq_0^*}^t \\
&= c_0 \|f\|_p^{p/p_0} \|g\|_{q^*}^{q^*/q_0^*}.
\end{aligned}$$

A similar calculation shows that

$$|G(1+iy)| \le c_1 \|f\|_p^{p/p_1} \|g\|_{q^*}^{q^*/q_1^*}.$$

We may now use the three lines lemma to deduce that

$$\begin{aligned}
|G(\lambda)| &\le c_0^{1-\lambda} c_1^{\lambda} \|f\|_p^{(1-\lambda)p/p_0 + \lambda p/p_1} \|g\|_{q^*}^{(1-\lambda)q^*/q_0^* + \lambda q^*/q_1^*} \\
&= c_0^{1-\lambda} c_1^{\lambda} \|f\|_p \|g\|_{q^*}.
\end{aligned}$$

The density of $\mathscr{D}$ in L^{q^*} implies that

$$\|T(f)\|_q \le c_0^{1-\lambda} c_1^{\lambda} \|f\|_p$$

for all $f \in \mathscr{D}$, from which the theorem follows easily. □

Corollary 3.6.3 *If $1 \le p \le 2$ and $1/p+1/q = 1$ then the Fourier transform $\mathcal{F}$ is a bounded operator from L^p to L^q.*

Proof We put $p_0 = 1$, $q_0 = \infty$, $p_1 = 2$ and $q_1 = 2$ into the above theorem. □

We now turn to some applications of the above results to the operator $H_0 := -\overline{\Delta}$.

Theorem 3.6.4 *If $0 < \alpha \le N/4$ and $2 \le p < 2N/(N-4\alpha)$, then $(H_0+1)^{-\alpha}$ is a bounded linear operator from L^2 to L^p. If $\alpha > N/4$, then $(H_0+1)^{-\alpha}$ is a bounded linear operator from L^2 to L^∞.*

Proof Assume first that $0 < \alpha \le N/4$. If we put $g(y) := (y^2+1)^{-\alpha}$ and assume that $N/2\alpha < s \le \infty$ then

$$\|g\|_s^s = c_1 \int_0^\infty (1+r^2)^{-\alpha s} r^{N-1} \mathrm{d}r < \infty.$$

If $k := \mathcal{F}(H_0+1)^{-\alpha} f$ and $f \in L^2$, then $k(y) = g(y)\mathcal{F}f(y)$ for all $y \in \mathbf{R}^N$. Putting $1/q := 1/s + 1/2$ we deduce that $1 < q \le 2$ and

$$\|k\|_q \le \|g\|_s \|\mathcal{F}f\|_2 = c_2 \|f\|_2.$$

If $1/p + 1/q = 1$ then $2 \le p < \infty$ and it follows from Corollary 3.6.3 that

$$\|(H_0+1)^{-\alpha} f\|_p = \|\mathcal{F}^{-1}k\|_p \le c_3 \|k\|_q \le c_4 \|f\|_2.$$

It is easy to check that p increases from 2 to $2N/(N-4\alpha)$ as s decreases from ∞ to $N/2\alpha$.

If $\alpha > N/4$ then the function g defined above lies in L^2 and we deduce that

$$\|k\|_1 \le \|g\|_2 \|\mathcal{F}f\|_2 = c_2 \|f\|_2.$$

The remainder of the proof is as before. □

Note The above theorem does not tell the full story as far as the extreme values of α and p are concerned. If $\alpha = N/4$, then $(H_0+1)^{-\alpha}$ is not a bounded operator from L^2 to L^∞. However, if $0 < \alpha < N/4$, then $(H_0+1)^{-\alpha}$ is bounded from L^2 to L^p for $p = 2N/(N-4\alpha)$. This fact is important for more advanced studies, but we shall not use it.

Theorem 3.6.5 *If $0 < \alpha \le N/4$ and $W \in L^q$ is a multiplication operator,* *then* $W(H_0+1)^{-\alpha}$ *is a bounded operator on* L^2 *provided* $N/2\alpha < q \le \infty$. *Moreover, there exists a constant* $c>0$ *such that*

$$\| W(H_0+1)^{-\alpha}\|_{L^2\to L^2} \le c\|W\|_q$$

for all such W. *The same bound holds if* $\alpha > N/4$ *and* $q = 2$. *In both cases we have*

$$\lim_{\lambda\to\infty} \| W(H_0+\lambda)^{-\alpha}\|_{L^2\to L^2} = 0.$$

Proof If $0<\alpha\le N/4$, then

$$\| W(H_0+1)^{-\alpha} f\|_2 \le \|W\|_q \|(H_0+1)^{-\alpha} f\|_p$$

provided $1/2 = 1/p + 1/q$. The condition $2 \le p < 2N/(N-4\alpha)$ is equivalent to $N/2\alpha < q \le \infty$. The case $\alpha > N/4$ is similar. The proof of the final assertion follows from the fact that if $g_\lambda(y) := (y^2+\lambda)^{-\alpha}$ then we have $\lim_{\lambda\to\infty}\|g_\lambda\|_s = 0$ in the proof of Theorem 3.6.4. □

Our corollary below should logically appear at the end of the next chapter, when the properties of compact operators have been discussed.

Corollary 3.6.6 *If* $0<\alpha\le N/4$ *and* $N/2\alpha < q < \infty$, *or* $\alpha > N/4$ *and* $q=2$, *then the operator* $W(H_0+1)^{-\alpha}$ *is actually a compact operator on* L^2.

Proof Let W_n be a sequence of functions in $L^2 \cap L^q$ which converges to W in L^q norm. Also let

$$g_n(y) := \begin{cases} (y^2+1)^{-\alpha} & \text{if } |y|\le n, \\ 0 & \text{otherwise.} \end{cases}$$

If we define $A_n : L^2 \to L^2$ by

$$A_n f := W_n \mathcal{F}^{-1}\{g_n(\mathcal{F} f)\},$$

then an examination of the steps of the above proofs shows that

$$\lim_{n\to\infty} \| W(H_0+1)^{-\alpha} - A_n\| = 0.$$

Since the set of compact operators is closed under norm limits, it is sufficient to prove that each A_n is a Hilbert–Schmidt operator; see Exercise 4.5 for the definition. Now A_n has the integral kernel

$$A_n(x,y) := cW_n(x)(\mathcal{F}^{-1}g_n)(x-y)$$

so

$$\begin{aligned}\|A_n\|_{HS}^2 &= c^2\|W_n\|_2^2\|\mathcal{F}^{-1}g_n\|_2^2 \\ &= c^2\|W_n\|_2^2\|g_n\|_2^2 < \infty.\end{aligned}$$

□

3.7 The Sobolev spaces $W^{n,2}(\mathbf{R}^N)$

The Sobolev spaces $\mathcal{H}^n := W^{n,2}(\mathbf{R}^N)$ were introduced by Sobolev in 1938. They are important because a number of operator estimates are naturally carried out using them. As n increases functions in $\mathcal{H}^n$ become better behaved locally, so it is advantageous to know that a function lies in $\mathcal{H}^n$ for a large value of n. We shall only use $\mathcal{H}^1$ and $\mathcal{H}^2$ in this text, and refer the reader to Adams (1975) for a more thorough account of the theory than is given here.

A function $f \in L^2$ is said to lie in $\mathcal{H}^n$ for a given positive integer n if the weak partial derivatives $D^\alpha f$ lie in L^2 for all $|\alpha| \le n$. We then define the Sobolev norm of such functions by

$$|||f|||_n^2 := \sum_{|\alpha|\le n} \|D^\alpha f\|_2^2.$$

Lemma 3.7.1 *A function $f \in L^2$ lies in $\mathcal{H}^n$ if and only if*

$$|||f|||_n^2 := \int_{\mathbf{R}^N} \Big(\sum_{|\alpha|\le n} |y^\alpha|^2\Big)|\mathcal{F}f(y)|^2 \mathrm{d}^N y < \infty.$$

The space $\mathcal{H}^n$ is a Hilbert space for the associated inner product.

Proof This is an immediate consequence of the formula (3.3.2) of Theorem 3.3.2. □

The definition can now be extended to non-positive and non-integral values of n. If $n \in \mathbf{R}$ we say that a distribution ϕ lies in $\mathcal{H}^n$ if $\mathcal{F}\phi$ is represented by a function g such that

$$\|\phi\|_n^2 := \int_{\mathbf{R}^N} (1+|y|^2)^n |g(y)|^2 \mathrm{d}^N y < \infty.$$

Note that we only have $\mathcal{H}^n \subseteq L^2$ if $n \ge 0$.

Lemma 3.7.2 *If n is a non-negative integer, then the two definitions are consistent and there exists a constant $c_n > 0$ such that*

$$c_n^{-1}\|f\|_n \le |||f|||_n \le c_n\|f\|_n$$

for all $f \in \mathcal{H}^n$.

Proof This follows from the inequality

$$c_n^{-1}(1+|y|^2)^n \le \sum_{|\alpha|\le n} |y^\alpha|^2 \le c_n(1+|y|^2)^n$$

valid for all $y \in \mathbf{R}^N$ provided c_n is large enough. □

Theorem 3.7.3 *The space $\mathcal{S}$ is norm dense in $\mathcal{H}^n$ for all $n \in \mathbf{R}$. Moreover, given $f \in \mathcal{S}$, there exists a sequence $f_m \in C_c^\infty$ such that*

$$\lim_{m\to\infty} \|f - f_m\|_n = 0$$

for all $n \in \mathbf{R}$.

Proof Given $f \in \mathcal{H}^n$ we define f_m by

$$\hat{f}_m(y) := \begin{cases} \hat{f}(y) & \text{if } |y| \le m, \\ 0 & \text{otherwise.} \end{cases}$$

It is easy to see that $f_m \in \mathcal{H}^n$ and $\lim_{m\to\infty} \|f - f_m\|_n = 0$. Therefore the set of functions in $\mathcal{H}^n$ whose Fourier transforms have compact support is norm dense in $\mathcal{H}^n$.

Now let $f \in \mathcal{H}^n$ have a Fourier transform of compact support and let f_s be defined for $s > 0$ by $\hat{f}_s := \hat{f} * k_s$, where k_s is a standard mollifier as constructed in Section 3.2. One sees as in the proof of Lemma 3.2.2 that $\hat{f}_s \in C_c^\infty$ and that $\lim_{s\to 0} \|f - f_s\|_n = 0$. Therefore the set of $f \in \mathcal{H}^n$ such that $\hat{f} \in C_c^\infty$ is dense in $\mathcal{H}^n$. This set is contained in $\mathcal{S}$, so $\mathcal{S}$ is also dense in $\mathcal{H}^n$.

We need only prove the second statement of the theorem for positive integers n, since norm convergence in $\mathcal{H}^n$ implies norm convergence in $\mathcal{H}^s$ for all $s \le n$. Let $\phi : \mathbf{R}^N \to [0,1]$ be a smooth function such that $\phi(x) = 1$ if $|x| \le 1$ and $\phi(x) = 0$ if $|x| \ge 2$. Given $f \in \mathcal{S}$, define $f_m \in C_c^\infty$ by

$$f_m(x) := f(x)\phi(x/m).$$

A direct calculation shows that $x^\beta D^\alpha f_m$ converges uniformly to $x^\beta D^\alpha f$ for all α and β. It follows that $D^\alpha f_m$ converges in L^2 norm to $D^\alpha f$ for all α, and this implies the second statement of the theorem for positive integers n. □

Theorem 3.7.4 *Let H be a uniformly elliptic constant coefficient differential operator of order n with domain C_c^∞ or $\mathscr{S}$. Then its closure $\overline{H}$ has domain equal to $\mathscr{H}^n$.*

Proof We assume that the domain is $\mathscr{S}$, the other case being handled using Theorem 3.7.3. The uniform ellipticity of H is equivalent to the assumption that its symbol $a(y)$ satisfies

$$c^{-1}(1+|y|^2)^{n/2} \le c_1 + a(y) \le c(1+|y|^2)^{n/2}$$

for some $c, c_1 > 0$ and all $y \in \mathbf{R}^N$. This implies

$$c^{-2}\|f\|_n^2 \le \int_{\mathbf{R}^N} |\{c_1 + a(y)\}\mathscr{F}f(y)|^2 \mathrm{d}^N y \le c^2\|f\|_n^2$$

for all $f \in \mathscr{S}$, or equivalently

$$c^{-1}\|f\|_n \le \|(c_1 + H)f\| \le c\|f\|_n$$

for all $f \in \mathscr{S}$. Since $\mathscr{S}$ is dense in $\mathscr{H}^n$, this establishes that $\mathscr{H}^n$ is the domain of the closure of H. □

Example 3.7.5 Let $H_0 := -\overline{\Delta}$, so that $\mathrm{Spec}(H_0) = [0, \infty)$ and $\mathrm{Dom}(H_0) = \mathscr{H}^2$. We may also write $H_0 = K^2$ where $K := (-\overline{\Delta})^{1/2}$ is the self-adjoint operator defined on $\mathscr{H}^1 = W^{1,2}(\mathbf{R}^N)$ by $\mathscr{F}Kf(y) := |y|\mathscr{F}f(y)$. Although K is not a differential operator, it is of substantial importance, as we shall see in Example 4.4.7. □

Theorem 3.7.6 *If $\phi \in \mathscr{H}^n$ for some $n > N/2$ then ϕ is (equal almost everywhere to) a bounded continuous function. If $\phi \in \mathscr{H}^n$ for all $n \in \mathbf{R}$ then $D^\alpha\phi$ are bounded smooth functions for all α.*

Proof If $\phi \in \mathscr{H}^n$, then

$$\Big\{\int_{\mathbf{R}^N} |\hat{\phi}(y)|\mathrm{d}^N y\Big\}^2 \le \int_{\mathbf{R}^N} (1+|y|^2)^n |\hat{\phi}(y)|^2 \mathrm{d}^N y \int_{\mathbf{R}^N} (1+|y|^2)^{-n} \mathrm{d}^N y$$
$$< \infty.$$

Since $\hat{\phi} \in L^1$, an application of the dominated convergence theorem establishes that

$$\phi(x) = (2\pi)^{-N/2} \int_{\mathbf{R}^N} \hat{\phi}(y) e^{ix\cdot y} d^N y$$

is bounded and continuous. If $\phi \in \mathcal{H}^n$ for all $n \in \mathbf{R}$ then $D^\alpha \phi \in \mathcal{H}^m$ for all α and all $m \in \mathbf{R}$. Therefore $D^\alpha \phi$ is bounded and continuous for all α. □

Exercises

3.1 The function $f : \mathbf{R}^N \to \mathbf{C}$ is given by

$$f(x) := \text{Poly}(x) \exp[-\alpha|x-a|^2]$$

where Poly is a polynomial in $x_1, ..., x_N$, $\text{Re}(\alpha) > 0$ and $a \in \mathbf{R}^N$. Prove that f lies in Schwartz space.

3.2 Prove that for all $k > 0$ there exist positive constants c_1, c_2 such that

$$c_1 \langle y \rangle^k \le 1 + \sum_{|\beta|=k} |y^\beta| \le c_2 \langle y \rangle^k$$

for all $y \in \mathbf{R}^N$. Use this to complete the proof of Theorem 3.3.2.

3.3 Let $f \in L^1$ and let k_s be the mollifier of Section 3.2. Prove that $k_s * f$ is a smooth function on $\mathbf{R}^N$ and that $D^\alpha(k_s * f) \in L^1$ for all multi-indexes α.

3.4 Let $\mathcal{E}$ denote the class of functions g on $\mathbf{R}^N$ such that

$$\int_{\mathbf{R}^N} |g(x)|(1+|x|)^{-m} d^N x < \infty$$

for some $m > 0$. Prove that $L^p(\mathbf{R}^N) \subseteq \mathcal{E}$ for all $1 \le p \le \infty$. Use (3.4.1) to associate a distribution in $\mathcal{S}'$ to any function $g \in \mathcal{E}$. (We comment that $\mathcal{E}$ is the largest class of functions with this property, if one imposes an extra continuity condition on the distribution.)

3.5 Prove that if $f \in L^2(\mathbf{R}^N)$ has compact support then its Fourier transform is smooth on $\mathbf{R}^N$. Indeed the Fourier transform can be extended to a holomorphic function defined on the whole of $\mathbf{C}^N$.

3.6 Prove that the equation $(-\Delta + 1)f = g$ has a weak solution $f \in L^2(\mathbf{R}^N)$ for all $g \in L^2(\mathbf{R}^N)$.

3.7 Prove that if $N \geq 5$ and $g \in \mathcal{S}$, then the equation $\Delta f = g$ has a solution $f \in L^2(\mathbf{R}^N)$.

3.8 Prove that if $N = 2$ and $g \in \mathcal{S}$ satisfies $g(x, y) = -g(-x, y) = -g(x, -y)$ for all $x, y \in \mathbf{R}$, then the equation $\Delta f = g$ has a solution $f \in L^2(\mathbf{R}^2)$.

3.9 Let H be the operator on $L^2(\mathbf{R}^2)$ associated with the expression

$$Hf := \frac{\partial^4 f}{\partial x^4} - \lambda \frac{\partial^2 f}{\partial y^2},$$

where $\lambda \in \mathbf{C}$. Find the spectrum of H for all possible complex values of λ.

3.10 Prove that if $N \geq 2$ and $2 \leq p < 2N/(N-2)$ then $W^{1,2}(\mathbf{R}^N) \subseteq L^p(\mathbf{R}^N)$.

3.11 Prove that the operator $H_0 := -\overline{\Delta}$ is bounded from its domain $\mathcal{H}^2$ to L^2, but also bounded as an operator from L^2 to $\mathcal{H}^{-2}$.

3.12 Prove that $(H_0 + 1)^{-1}$ may be regarded as a bounded invertible linear operator from $\mathcal{H}^n$ onto $\mathcal{H}^{n+2}$ for all $n \in \mathbf{R}$.

3.13 An operator $A : C_c^\infty \to L^2$ is said to be local if the support of Af is contained in the support of f for all $f \in C_c^\infty$. Use Exercise 3.5 to prove that the operator K of Example 3.7.5 is not local.

3.14 Prove that if $k \in L^1(\mathbf{R}^N)$ and $1 \leq p \leq \infty$ then the formula $T(f) := k * f$ defines a bounded linear operator T on $L^p(\mathbf{R}^N)$ with norm at most $\|k\|_1$. Hint: Interpolate between the cases $p = 1$ and $p = \infty$.

4

The variational method

4.1 Classification of the spectrum

In this chapter we investigate the spectral properties of abstract self-adjoint operators in more detail, making use of the spectral theorems (Theorems 2.3.1 and 2.5.1). In contrast with Chapter 3, we focus mainly on abstract situations in which the spectrum of a self-adjoint operator consists simply of isolated eigenvalues of finite multiplicity. Several classes of operators of this type are studied in later chapters. We use the notation of Theorem 2.5.1 freely within this section, but frequently suppress explicit reference to the unitary operator U of that theorem. Many of the considerations of this chapter become trivial or do not make sense if the Hilbert space $\mathcal{H}$ is finite-dimensional, so we assume throughout that it is infinite-dimensional. We assume as always that $\mathcal{H}$ is separable, or equivalently that any complete orthonormal set is countable.

The following lemma from measure theory will be important.

Lemma 4.1.1 *Let μ be a finite measure on $\mathbf{R}^N$, and let E be a Borel subset of $\mathbf{R}^N$. Then the space $L := L^2(E, \mathrm{d}\mu)$ has finite dimension m if and only if there are m distinct points $x_1, ..., x_m$ in E of positive measure such that $E \backslash \cup_{r=1}^{m} \{x_r\}$ has zero measure.*

Proof If such points exist then it is clear that L has a basis consisting of the characteristic functions of these points. Conversely suppose that L has finite dimension m. For any integer $r \geq 1$ we can partition $\mathbf{R}^N$ into a countable number of cubes $C(r,s), 1 \leq s < \infty$, with edge length 2^{-r}; we take these cubes to be in standard position, that is every coordinate of every vertex of $C(r,s)$ is of the form $t/2^r$ where t is an integer; we also assign faces to cubes in any manner which ensures that the cubes are disjoint and have union equal to $\mathbf{R}^N$. At most m of the sets $E \cap C(r,s), 1 \le s < \infty$, can be non-trivial in the sense of having non-zero measures. As r increases these non-trivial sets are nested downwards, and cannot increase in number beyond m. Therefore they converge onto at most m points, whose complement has zero measure. By the first part of the theorem the number of points is exactly m. □

Lemma 4.1.2 *A number $\lambda \in \mathbf{R}$ lies in the spectrum* Spec(H) *of a self-adjoint operator H if and only if there exists a sequence of functions $f_n \in$* Dom(H) *with $\|f_n\| = 1$ such that*

$$\lim_{n\to\infty} \|Hf_n - \lambda f_n\| = 0.$$

Proof If a sequence f_n with the above property exists, then the inverse operator $(H-\lambda)^{-1}$, if it exists, cannot be bounded. Therefore $\lambda \in \text{Spec}(H)$. Conversely if $\lambda \in \text{Spec}(H)$, then λ lies in the essential range of the function h of Corollary 2.5.4, and a suitable sequence exists by the proof of Lemma 1.3.2. □

It is entirely possible that no point of the spectrum of H is an eigenvalue. In the spectral representation of Theorem 2.5.1, a number $\lambda \in \mathbf{R}$ is an eigenvalue of H if and only if the set

$$E := \{(\lambda, n) : n \in \mathbf{N}\}$$

has non-zero measure. If this happens then the eigenspace associated with the eigenvalue is unitarily equivalent to $L^2(E, \mathrm{d}\mu)$; in particular these two linear spaces have the same dimension.

The point spectrum of H is by definition just the set of all of its eigenvalues. The discrete spectrum, however, is defined as the set of all eigenvalues λ of finite multiplicity which are isolated in the sense that $(\lambda-\varepsilon, \lambda)$ and $(\lambda, \lambda+\varepsilon)$ are disjoint from the spectrum for some $\varepsilon > 0$. The non-discrete part of the spectrum of H is called the essential spectrum, and is denoted by EssSpec(H).

Example 4.1.3 Let $H_0 = -\Delta$ with dense domain C_c^∞ in $\mathscr{H} = L^2(\mathbf{R}^N, \mathrm{d}x)$. Since we have already seen in Chapter 3 that the spectrum of $\overline{H}_0$ equals $[0, \infty)$, it follows that the spectrum of $\overline{H}_0$ equals its essential spectrum. Indeed $\overline{H}_0$ has no point spectrum, since the relevant set

$\{y \in \mathbf{R}^N : |y|^2 = \lambda\}$ has zero Lebesgue measure for all $\lambda \in \mathbf{R}$. Similar conclusions can be drawn for every constant coefficient differential operator on $L^2(\mathbf{R}^N, \mathrm{d}x)$. □

We refer the reader to the end of Section 2.5 for the definition of spectral subspace used in the following lemma.

Lemma 4.1.4 *The essential spectrum is a closed subset of* Spec(H). *A number λ lies in the essential spectrum of H if and only if the spectral subspace*

$$\begin{aligned} L_{(\lambda-\varepsilon,\lambda+\varepsilon)} &= P_{(\lambda-\varepsilon,\lambda+\varepsilon)}\mathcal{H} \\ &= L^2(E_{(\lambda-\varepsilon,\lambda+\varepsilon)}, \mathrm{d}\mu) \end{aligned}$$

is infinite-dimensional for all $\varepsilon > 0$.

Proof Each point of the discrete spectrum is an isolated point in Spec(H), so the discrete spectrum is a relatively open subset of the whole spectrum. Assume that $\lambda \in$ Spec(H) and that the spectral subspace $L^2(E_{(\lambda-\varepsilon,\lambda+\varepsilon)}, \mathrm{d}\mu)$ is finite-dimensional for some $\varepsilon > 0$. By Lemma 4.1.1 there exist $x_1, ..., x_m \in E_{(\lambda-\varepsilon,\lambda+\varepsilon)}$ which have positive measure and whose complement has zero measure. If $x_r = (s_r, n_r)$, then s_r are eigenvalues of finite multiplicity, and every other number in $(\lambda - \varepsilon, \lambda + \varepsilon)$ lies outside the spectrum. This establishes that $\lambda = s_r$ for some r and that λ lies in the discrete spectrum. The converse argument is similar but easier. □

Theorem 4.1.5 *The essential spectrum of a self-adjoint operator H is empty if and only if there is a complete orthonormal set of eigenfunctions $\{f_n\}_{n=1}^{\infty}$ of H such that the corresponding eigenvalues λ_n converge in absolute value to ∞ as $n \to \infty$.*

Proof If the essential spectrum is empty then the spectrum consists of a set $\{s_r\}_{r=1}^{\infty}$ of isolated eigenvalues, which can only converge to $\pm\infty$. Each eigenvalue has finite multiplicity $m_r < \infty$. If we enumerate the eigenvalues in order of increasing absolute values and repeat each eigenvalue according to its multiplicity, then there exists an associated orthonormal set of eigenfunctions $\{f_n\}$. Suppose that this set is not complete. Then the linear subspace

$$L := \{f \in \mathcal{H} : \langle f, f_n \rangle = 0 \text{ for all } n \geq 1\}$$

is invariant with respect to H in the sense of Section 2.4. The spectrum of

H restricted to this subspace is non-empty by Theorem 1.2.10. Therefore H has further eigenvalues and eigenfunctions not accounted for in the above list.

Conversely suppose that a sequence of eigenvalues and eigenfunctions with the stated properties exists, and let $\{s_r\}_{r=1}^{\infty}$ be the set of distinct eigenvalues. By the assumption that λ_n converge in absolute value to ∞ as $n \to \infty$, we deduce that s_r are isolated eigenvalues of finite multiplicity. It follows from Lemma 1.2.2 that $\mathrm{Spec}(H) = \bigcup_{r=1}^{\infty}\{s_r\}$. Thus H has no essential spectrum. □

4.2 Compact operators

The spectral theory of compact self-adjoint operators can be developed as a special case of the general theory of compact operators on Banach spaces, due to Riesz (1918). We do not adopt this approach here, but show how to derive the results from the spectral theory of Chapter 2.

In what ensues we shall assume that the reader knows that the following three properties of a complete metric space M with metric $d(x, y)$ are all equivalent:

(1) Every sequence in M has a convergent subsequence.
(2) M is compact, that is every open covering has a finite subcovering.
(3) M is totally bounded, that is for every $\varepsilon > 0$ there exists a finite set $x_1, ..., x_s$ in M such that every $x \in M$ satisfies $d(x_t, x) < \varepsilon$ for some t, $1 \le t \le s$.

We define a compact operator on a Banach space $\mathscr{B}$ to be an operator A with domain $\mathscr{B}$ such that for any bounded sequence $f_n \in \mathscr{B}$, Af_n has a norm convergent subsequence. Clearly any compact operator is bounded. Any bounded operator of finite rank (that is with range space of finite dimension) is compact. The following lemma can be generalized to many of the commonly used Banach spaces by constructing an appropriate sequence of projections. We show in Exercise 4.4 that it may be used to prove many of the standard algebraic properties of the set of all compact operators.

Lemma 4.2.1 *An operator A on a Hilbert space $\mathscr{H}$ is compact if and only if there exists a sequence of operators A_n of finite rank which converge in norm to A.*

Proof If B denotes the unit ball in $\mathcal{H}$, D is the closure of $A(B)$ and A is compact, then D is a norm compact subset of $\mathcal{H}$. If $\{\phi_n\}_{n=1}^{\infty}$ is a complete orthonormal set in $\mathcal{H}$, then the projections P_n defined by

$$P_n f := \sum_{r=1}^{n} \langle f, \phi_r \rangle \phi_r \tag{4.2.1}$$

are all of finite rank, so the operators $P_n A$ are also of finite rank. In order to prove that $P_n A$ converge in norm to A it is sufficient to prove that $1 - P_n$ converges uniformly to zero on the compact set D.

The total boundedness of D implies that for any $\varepsilon > 0$ there exists a finite set $\{x_1, ..., x_s\}$ in D such that $x \in D$ implies $\|x - x_t\| < \varepsilon/2$ for some t such that $1 \leq t \leq s$. It follows from (4.2.1) that $\lim_{n\to\infty}\{P_n f - f\} = 0$ for all $f \in \mathcal{H}$. Therefore there exists N such that $n \geq N$ implies $\|(1 - P_n)x_t\| < \varepsilon/2$ for all $1 \leq t \leq s$. If $n \geq N$ and $x \in D$, then

$$\begin{aligned} \|(1 - P_n)x\| &< \|(1 - P_n)x_t\| + \|(1 - P_n)(x_t - x)\| \\ &\leq \varepsilon/2 + \varepsilon/2 = \varepsilon. \end{aligned}$$

This establishes that $1 - P_n$ converges uniformly to 0 on D.

For the converse part of the lemma suppose that A is bounded and that there exists a sequence A_n of finite rank operators which converges in norm to A. Given $\varepsilon > 0$ fix n so that $\|A - A_n\| < \varepsilon/2$. Since A_n is of finite rank and hence compact, there exists a finite set $x_1, ..., x_s \in \mathcal{H}$ such that $x \in B$ implies $\|A_n x - x_t\| < \varepsilon/2$ for some t. It follows that for every $x \in B$ we have $\|Ax - x_t\| < \varepsilon$ for some t. Hence the set $A(B)$ is totally bounded and its closure is compact; in other words A is a compact operator. □

Theorem 4.2.2 *If H is a compact self-adjoint operator on a Hilbert space $\mathcal{H}$, then there is a complete orthonormal set of eigenvectors $\{\phi_n\}_{n=1}^{\infty}$ of H with corresponding eigenvalues λ_n which converge to 0 as $n \to \infty$. In particular any non-zero eigenvalue of H is of finite multiplicity.*

Proof By Theorem 2.5.1 there is a unitary operator U from $\mathcal{H}$ to $L^2 = L^2(S \times \mathbf{N}, \mathrm{d}\mu)$ such that H is unitarily equivalent to the operator of multiplication by $h(s, n) = s$ on L^2. In the following proof we suppress explicit reference to U. For any integer $n \geq 0$ let E_n denote the set of points (s, n) such that $2^{-n-1}\|H\| < |h(s, n)| \leq 2^{-n}\|H\|$.

We first claim that $L^2(E_n)$ is finite-dimensional for every n. If this were not true then there would exist an infinite orthonormal set $\{f_r\}_{r=1}^{\infty}$ in this subspace. If $r \neq p$ then considering H as a multiplication operator we see that

$$\|H(f_r - f_p)\|^2 = \int_{E_n} h(s,n)^2 |f_r(s,n) - f_p(s,n)|^2 \mathrm{d}\mu$$
$$\geq 2^{-2n-2}\|H\|^2 \int_{E_n} |f_r(s,n) - f_p(s,n)|^2 \mathrm{d}\mu$$
$$\geq 2^{-2n-2}\|H\|^2,$$

which is not compatible with the condition that $\{Hf_r\}_{r=1}^{\infty}$ must have a norm convergent subsequence.

It now follows by Lemma 4.1.1 that the operator H has finite spectrum in the intervals $(2^{-n-1}\|H\|, 2^{-n}\|H\|]$ and $[-2^{-n}\|H\|, -2^{-n-1}\|H\|)$ and that each spectral point in these intervals is an eigenvalue of finite multiplicity. We can therefore obtain a finite orthonormal basis of $L^2(E_n)$ consisting of eigenvectors of H. The theorem follows by putting together all these orthonormal sets and also a possibly infinite orthonormal basis of the subspace $L^2(E)$, where E is the set of points where $h = 0$. □

The constant 1 in part (1) of the following corollary has no particular status; see Exercise 4.9.

Corollary 4.2.3 *Let H be an unbounded self-adjoint operator on the Hilbert space $\mathscr{H}$ which is non-negative in the sense that* Spec$(H) \subseteq [0, \infty)$. *Then the following conditions are equivalent:*

(1) *The resolvent operator $(H + 1)^{-1}$ is compact.*
(2) *The operator H has empty essential spectrum.*
(3) *There exists a complete orthonormal set of eigenvectors $\{\phi_n\}_{n=1}^{\infty}$ of H with corresponding eigenvalues $\lambda_n \geq 0$ which converge to $+\infty$ as $n \to \infty$.*

Proof (1)⇒(3) According to Theorem 2.5.1, H is unitarily equivalent to multiplication by the unbounded non-negative function $h(s,n) = s$ on a space $L^2 = L^2(S \times \mathbf{N}, \mathrm{d}\mu)$. Also $(H + 1)^{-1}$ is unitarily equivalent to multiplication by $k(s,n) = (h(s,n) + 1)^{-1}$ by Lemma 1.3.2. The proof of Theorem 4.2.2 yields (3) at the same time as it yields the spectral behaviour of the resolvent.

(3)⇒(1) We start by rearranging the eigenvectors so that the sequence $\{\lambda_n\}_{n=1}^{\infty}$ is non-decreasing. Define the finite rank operators A_n by

$$A_n f := \sum_{r=1}^{n} (\lambda_r + 1)^{-1} \langle f, \phi_r \rangle \phi_r.$$

From the formula

$$\{(H+1)^{-1} - A_n\}f = \sum_{r=n+1}^{\infty} (\lambda_r + 1)^{-1} \langle f, \phi_r \rangle \phi_r$$

we deduce that

$$\|\{(H+1)^{-1} - A_n\}f\| \le (\lambda_n + 1)^{-1} \|f\|$$

from which we see that A_n converge in norm to $(H+1)^{-1}$ as $n \to \infty$. The proof is completed by the use of Lemma 4.2.1.

(2)⇔(3) This is a special case of Theorem 4.1.5. □

Note A further important necessary and sufficient condition for the compactness of $(H+1)^{-1}$ is given in Exercise 4.2.

Example 4.2.4 It follows from Corollary 4.2.3 that the operators $\overline{H}_D$ and $\overline{H}_N$ defined in Example 1.1.1 both have compact resolvents. We shall see later that this is typical for second order elliptic operators defined on $L^2(\Omega, \mathrm{d}^N x)$ where Ω is a bounded region in $\mathbf{R}^N$. □

4.3 Positivity and fractional powers

Many of the partial differential operators which we shall study are semibounded or non-negative. We say that a symmetric operator H with domain $\mathcal{D}$ is non-negative if

$$\langle Hf, f \rangle \ge 0$$

for all $f \in \mathcal{D}$. If H is self-adjoint there are several equivalent ways of defining the notion. Non-negativity corresponds to the case $c = 0$ of the following theorem, and in the general case we write $H \ge c$, and say that H is semibounded.

Theorem 4.3.1 *Let H be a self-adjoint operator on the Hilbert space $\mathcal{H}$ and let $c \in \mathbf{R}$. Then the following conditions are equivalent:*

(1) *One has $\langle Hf, f \rangle \ge c\|f\|^2$ for all $f \in \mathrm{Dom}(H)$.*
(2) *The spectrum of H is contained in $[c, \infty)$.*
(3) *The function $(s, n) \to h(s, n) - c$ on $S \times \mathbf{N}$ is non-negative except on a set of zero μ-measure.*

Proof This is all immediate from Theorem 2.5.1 or Corollary 2.5.4. The fact that the L^2 representation of these theorems is not canonically determined by the operator is no problem. □

Example 4.3.2 The operator H_N of Example 1.1.1 on $L^2(a,b)$ is non-negative by virtue of the fundamental equality

$$\int_a^b (-f'')\overline{f}\mathrm{d}x = \int_a^b |f'|^2 \mathrm{d}x \geq 0,$$

proved by integration by parts. This inequality extends to the closure of the operator, which is therefore a non-negative self-adjoint operator by Example 1.2.3. The operator H_D is also non-negative but we may even take c to be the smallest eigenvalue $\pi^2/(b-a)^2$ of H_D in Theorem 4.3.1. □

Theorem 4.3.1 states that a non-negative self-adjoint operator H is unitarily equivalent to a non-negative multiplication operator h on an L^2 space. Given any $\lambda > 0$ we would like to define the unbounded self-adjoint operator H^λ so that it is unitarily equivalent to the multiplication operator h^λ. The only possible danger in doing this, that the above representation is not canonical, is easily resolved.

Theorem 4.3.3 *The operator H^λ on $\mathcal{H}$ defined in the above fashion is canonically determined by the functional relation*

$$(H^\lambda + 1)^{-1} = f(H),$$

where the continuous function f on $\mathbf{R}$ is defined by

$$f(x) := (|x|^\lambda + 1)^{-1} \tag{4.3.1}$$

and we are using the functional calculus of Theorem 2.3.1. If $H \geq a > 0$ and $0 < \lambda < 1$, then we also have the identity

$$H^{-\lambda} = c\int_0^\infty t^{-\lambda}(H+t)^{-1}\mathrm{d}t, \tag{4.3.2}$$

where the integral above is convergent in norm, and the constant $c > 0$ is given by

$$c^{-1} := \int_0^\infty t^{-\lambda}(1+t)^{-1}\mathrm{d}t.$$

Proof An application of Theorem 2.3.1 establishes that there is a canonically determined bounded self-adjoint operator $f(H)$ associated with

(4.3.1). Theorem 2.5.1 then shows that $f(H)$ is one-one, so there is a canonical (and generally unbounded) self-adjoint operator which we call K such that $(K+1)^{-1} = f(H)$. Theorem 2.5.1 then also establishes that K is given by multiplication by h^{λ}.

The norm continuity of the integrand of (4.3.2) is proved in Lemma 1.1.2. The convergence of the integral follows from Lemma 1.3.2 and Corollary 2.5.4. If $\phi, \psi \in \mathcal{H}$ and A denotes the operator defined by the right hand side of (4.3.2) then the use of Fubini's theorem yields

$$\begin{aligned}\langle A\phi, \psi\rangle &= c\int_0^{\infty}\int_{S\times\mathbf{N}} t^{-\lambda}\{h(s,n)+t\}^{-1}\phi(s,n)\overline{\psi(s,n)}\mathrm{d}\mu(s,n)\mathrm{d}t \\ &= \int_{S\times\mathbf{N}} h(s,n)^{-\lambda}\phi(s,n)\overline{\psi(s,n)}\mathrm{d}\mu(s,n) \\ &= \langle H^{-\lambda}\phi, \psi\rangle.\end{aligned}$$

Since ϕ and ψ are arbitrary, this establishes the second statement of the theorem. □

The fractional powers H^{λ} of a non-negative self-adjoint differential operator H are hard to write down explicitly. However, we shall see that $\mathrm{Dom}(H^{1/2})$ may be much simpler to describe than $\mathrm{Dom}(H)$. The case $\lambda = \frac{1}{2}$ of the following theorem is of particular importance and is reformulated in Lemma 4.4.1 below.

Theorem 4.3.4 *If H is a non-negative self-adjoint operator and $0 < \lambda < 1$, then $f \in \mathrm{Dom}(H)$ if and only if $f \in \mathrm{Dom}(H^{\lambda})$ and also $H^{\lambda}f \in \mathrm{Dom}(H^{1-\lambda})$. For such f we have*

$$Hf = H^{1-\lambda}(H^{\lambda}f).$$

Proof We follow the notation of Theorem 2.5.1, but suppress explicit reference to the unitary operator U. The theorem then becomes the elementary statement that

$$\int (1+|h|^2)|f|^2\mathrm{d}\mu < \infty$$

if and only if we have both

$$\int (1+|h|^{2\lambda})|f|^2\mathrm{d}\mu < \infty$$

and

$$\int (1+|h|^{2-2\lambda})|h^{\lambda}f|^2\mathrm{d}\mu < \infty.$$

□

4.4 Closed quadratic forms

In this section and the next we describe the two most important ideas in the whole book as far as applications of the spectral theorem to differential operators are concerned. The first is that it is often better to study the quadratic form associated with a non-negative self-adjoint operator H rather than the operator itself. Given a non-negative self-adjoint operator H we define $Q'(f,g)$ for $f,g \in \mathrm{Dom}(H^{1/2})$ by

$$Q'(f,g) := \langle H^{1/2}f, H^{1/2}g \rangle.$$

Experience over the last twenty years has shown that quadratic forms provide an extremely powerful tool for studying second order elliptic differential operators. Perhaps the strongest single reason for this is that many such operators with quite different domains have quadratic forms with the same domain. The following is a first step in a programme of rebasing the subject upon Q' and $\mathrm{Dom}(H^{1/2})$ instead of H and $\mathrm{Dom}(H)$.

Lemma 4.4.1 *Let H be a non-negative self-adjoint operator on $\mathscr{H}$. Then $f \in \mathscr{H}$ lies in* $\mathrm{Dom}(H)$ *if and only if $f \in$* $\mathrm{Dom}(H^{1/2})$ *and also there exists $k \in \mathscr{H}$ such that*

$$Q'(f,g) = \langle k, g \rangle \tag{4.4.1}$$

for all $g \in$ $\mathrm{Dom}(H^{1/2})$. *In this case we have $Hf = k$.*

Proof The equation (4.4.1) is equivalent to the condition that $H^{1/2}f \in \mathrm{Dom}\{(H^{1/2})^*\}$ and $(H^{1/2})^* H^{1/2} f = k$. Since $(H^{1/2})^* = H^{1/2}$ the lemma is equivalent to the case $\lambda = \frac{1}{2}$ of Theorem 4.3.4. □

Our next goal is to characterise the quadratic forms which arise from non-negative self-adjoint operators in the above manner. We define a (non-negative) sesquilinear form Q' on a dense domain $\mathscr{D}$ in a Hilbert space $\mathscr{H}$ to be a map $Q' : \mathscr{D} \times \mathscr{D} \to \mathbf{C}$ such that

(1) $Q'(f,g)$ is linear in f.
(2) $Q'(f,g)$ is conjugate linear in g.
(3) $Q'(f,g) = \overline{Q'(g,f)}$ for all $f,g \in \mathscr{D}$.
(4) $Q'(f,f) \geq 0$ for all $f \in \mathscr{D}$.

We now define the quadratic form $Q : \mathscr{H} \to [0,+\infty]$ associated with Q' to be

$$Q(f) := \begin{cases} Q'(f,f) & \text{if } f \in \mathscr{D}, \\ +\infty & \text{otherwise.} \end{cases}$$

In this section we shall always distinguish between Q and Q', but elsewhere we shall leave the reader to work out which is intended from the context. The considerations of this section will be generalized to semibounded operators and forms in Section 8.1.

The following theorem uses the notion of lower semicontinuity (lsc). A function $f : M \to (-\infty, +\infty]$ on a metric space M is said to be lsc if for every convergent sequence $x_n \to x$ in M we have

$$f(x) \leq \liminf_{n\to\infty} f(x_n).$$

It is easy to prove that the pointwise limit of an increasing sequence of continuous functions is a lsc function.

Theorem 4.4.2 *The following conditions are equivalent:*

(1) *Q is the form arising from a non-negative self-adjoint operator H.*
(2) *Q is a lower semicontinuous function on $\mathscr{H}$.*
(3) *The domain $\mathscr{D}$ of Q is complete for the norm defined by*

$$|||f||| := \left(Q(f) + \|f\|^2\right)^{1/2}. \tag{4.4.2}$$

Proof We first note that $|||\ |||$ is the norm on $\mathscr{D}$ associated with the inner product

$$\langle f, g\rangle_1 := Q'(f,g) + \langle f, g\rangle.$$

(1)⇒(2) If we define $Q_n : \mathscr{H} \to [0,\infty)$ by

$$Q_n(f) := \langle nH(n+H)^{-1} f, f\rangle$$

then Q_n is bounded on the unit ball of $\mathscr{H}$ and continuous. By using the spectral theorem, Theorem 2.5.1, we see that $Q_n(f)$ increases monotonically to $Q(f)$ for every $f \in \mathscr{H}$. This implies that Q is lower semicontinuous.

(2)⇒(3) Let $\{f_n\}_{n=1}^\infty$ be a Cauchy sequence with respect to $|||\ |||$. Then $\{f_n\}_{n=1}^\infty$ is also Cauchy with respect to $\|\ \|$ and therefore converges to a limit $f \in \mathscr{H}$. Given $\varepsilon > 0$ there exists N such that

$$Q(f_m - f_n) + \|f_m - f_n\|^2 < \varepsilon^2$$

for all $m, n > N$. Letting $m \to \infty$ and using the lower semicontinuity of Q, we deduce that $f \in \mathscr{D}$ and

$$Q(f - f_n) + \|f - f_n\|^2 \leq \varepsilon^2$$

for all $n > N$. Thus $|||f - f_n||| \leq \varepsilon$, and we have shown that $\mathscr{D}$ is complete for the norm $|||\ |||$.

(3)⇒(1) We define $\mathscr{H}'$ to be the linear space $\mathscr{D}$ provided with the inner product $\langle\ ,\ \rangle_1$. If we consider $\langle\ ,\ \rangle$ to be a bounded form on $\mathscr{H}'$ then we discover that there exists a non-negative, self-adjoint contraction A on $\mathscr{H}'$ such that

$$\langle f, g\rangle = \langle Af, g\rangle_1$$

for all $f, g \in \mathscr{H}'$. By Theorem 2.5.1 there exist a unitary operator U from $\mathscr{H}'$ to $L^2(S \times \mathbf{N}, \mathrm{d}\mu')$, where $S \subseteq [0, 1]$, and a function a on $S \times \mathbf{N}$ satisfying $0 \leq a(s, n) \leq 1$ for all (s, n), such that A is unitarily equivalent to multiplication by a. Since $\langle Af, f\rangle_1 = 0$ implies $f = 0$ we see that $a(s, n) > 0$ except on a set of zero μ'-measure. Therefore we can write

$$a(s, n) = \{h(s, n) + 1\}^{-1},$$

where $h(s, n) \geq 0$ for all $(s, n) \in S \times \mathbf{N}$. We have now proved that

$$\langle f, g\rangle = \int_{S\times\mathbf{N}} f(s, n)\overline{g(s, n)}\{h(s, n) + 1\}^{-1}\mathrm{d}\mu'$$

and

$$Q'(f, g) + \langle f, g\rangle = \int_{S\times\mathbf{N}} f(s, n)\overline{g(s, n)}\,\mathrm{d}\mu'$$

for all $f, g \in \mathscr{D}$. If we define the new measure μ by

$$\mathrm{d}\mu(s, n) := \{h(s, n) + 1\}^{-1}\mathrm{d}\mu'(s, n),$$

then the above two formulae imply that $\mathscr{H}$ is unitarily equivalent to $L^2(S \times \mathbf{N}, \mathrm{d}\mu)$ and that

$$Q'(f, g) = \int_{S\times\mathbf{N}} f(s, n)\overline{g(s, n)}h(s, n)\,\mathrm{d}\mu$$

for all $f, g \in \mathscr{D}$. If H is the non-negative self-adjoint operator on $\mathscr{H}$ unitarily equivalent to the multiplication operator h on $L^2(S \times \mathbf{N}, \mathrm{d}\mu)$, it follows that Q is the quadratic form associated with H. □

We say that forms satisfying the conditions of the above theorem are closed. A form Q_2 is said to be an extension of Q_1 if it has a larger domain but coincides with Q_1 on the domain of Q_1. A form Q is said to be closable if it has a closed extension, and the smallest closed extension is called its closure $\overline{Q}$. A linear subspace $\mathscr{D}$ of the domain of a closed form Q is called a core for Q if Q is the closure of its restriction to $\mathscr{D}$. Note that the domain of $\overline{Q}$ is the completion of the domain of Q for the norm (4.4.2); this completion is definable for all forms Q but is not always embeddable as a subspace of $\mathcal{H}$.

One has to be as careful about specifying the domain of a quadratic form as one does for an operator. If for example one considers $H_0 = -\Delta$ acting on the domain $C_c^\infty \subseteq L^2(\mathbf{R}^N)$, then it is clear that

$$Q'(f,g) = \langle H_0 f, g\rangle = \int_{\mathbf{R}^N} \nabla f \cdot \overline{\nabla g}\, \mathrm{d}x$$

for all $f, g \in C_c^\infty$. The proof that Q is closable and the identification of the precise domain of its closure are issues which we will address below.

Corollary 4.4.3 *Let Q_1 and Q_2 be two quadratic forms on $\mathcal{H}$ with the same dense domain $\mathcal{D}$, and suppose that there exists a positive constant c such that*

$$c^{-1}Q_1(f) \leq Q_2(f) \leq cQ_1(f)$$

for all $f \in \mathcal{D}$. If Q_1 is the form of a non-negative self-adjoint operator H_1, then Q_2 is also the form of a non-negative self-adjoint operator H_2. Moreover,

$$\mathrm{Dom}(H_1^{1/2}) = \mathrm{Dom}(H_2^{1/2}) = \mathcal{D}.$$

The domains of the operators themselves may, however, be unrelated.

Proof By Theorem 4.4.2 the space $\mathcal{D}$ is complete with respect to the norm $|||\ |||_1$. The hypothesis of the corollary implies that $\mathcal{D}$ is also complete with respect to the equivalent norm $|||\ |||_2$, and the main part of the corollary follows by a second application of Theorem 4.4.2. The final statement is dealt with in the next example. □

Example 4.4.4 Consider the operator H on $L^2((\alpha,\beta), \mathrm{d}x)$ defined by

$$Hf(x) := -\frac{\mathrm{d}}{\mathrm{d}x}\left\{a(x)\frac{\mathrm{d}f}{\mathrm{d}x}\right\}$$

subject to Dirichlet boundary conditions. We assume that $c^{-1} < a(x) < c$ for some positive constant c and all $x \in [\alpha,\beta]$, and that $a(x)$ is smooth except for a finite number of simple jumps. We shall see that the discontinuities of $a(x)$ are of importance for the definition of the domain of the operator H but not for the domain of $\overline{H}^{1/2}$. For similar examples in higher dimensions we shall not even attempt to define the operator directly, but will work entirely through Theorem 4.4.2 or Corollary 4.4.3.

We define the domain $\mathscr{D}$ of H to be the set of all continuous and piecewise continuously differentiable functions f on $[\alpha, \beta]$ such that $f(\alpha) = f(\beta) = 0$; we also impose the condition on functions in $\mathscr{D}$ that $a(x)f'(x)$ should be continuously differentiable on $[\alpha, \beta]$. Integration by parts establishes that the quadratic form Q of H is given by

$$Q(f) = \int_\alpha^\beta a(x)|f'(x)|^2 \mathrm{d}x. \tag{4.4.3}$$

We now define a larger domain on which the quadratic form still makes sense.

Let $W_0^{1,2}(\alpha, \beta)$ denote the space of all functions f on $[\alpha, \beta]$ of the form

$$f(x) = \int_\alpha^x g(s)\mathrm{d}s,$$

where g is a function in $L^2(\alpha, \beta)$ such that $\int_\alpha^\beta g(s)\mathrm{d}s = 0$. Such functions f are continuous and vanish at α and β. We write $f' := g$ although this is not to be interpreted in the sense of pointwise derivatives. This definition is shown to be equivalent to a more general definition of $W_0^{1,2}$ in Lemma 7.1.1. If we define the form $\tilde{Q}$ with domain $W_0^{1,2}(\alpha, \beta)$ by (4.4.3) above, then it is not hard to show that $\tilde{Q}$ is a closed extension of Q.

We now show that $\tilde{Q}$ is actually the closure of Q. The importance of this fact is that, unlike the domain of the operator H, the space $W_0^{1,2}(\alpha, \beta)$ is independent of the discontinuities of the coefficient function $a(x)$. If $h \in L^2(\alpha, \beta)$ and

$$\int_\alpha^\beta h(s)a(s)^{-1}\mathrm{d}s = 0,$$

there exists a sequence of polynomials h_n which converge to h in L^2 norm and satisfy

$$\int_\alpha^\beta h_n(s)a(s)^{-1}\mathrm{d}s = 0$$

for all $n \geq 1$. If we put

$$f_n(x) := \int_\alpha^x h_n(s)a(s)^{-1}\mathrm{d}s,$$

then it is easy to check that $f_n \in \mathrm{Dom}(H)$ and that $|||f_n - f||| \to 0$ as $n \to \infty$. □

The use of quadratic forms allows us to prove that every non-negative symmetric operator H has at least one non-negative self-adjoint extension. If H is not essentially self-adjoint then this extension, called the Friedrichs extension, is one of the infinitely many possible self-adjoint extensions. Although Friedrichs proved the following theorem in 1934, the formulation of the more general Theorem 4.4.2, upon which our proof depends, was delayed until a paper of Lax and Milgram in 1954.

Theorem 4.4.5 *Let Q' be the form defined on the domain $\mathcal{D}$ of a non-negative symmetric operator H by*

$$Q'(f,g) := \langle Hf, g\rangle.$$

Then Q is closable and its closure is associated with a self-adjoint extension of H. It follows that H has equal deficiency indices.

Proof Let $\mathcal{H}_1$ be the abstract completion of the domain $\mathcal{D}$ of H for the norm $|||\ |||$ defined in Theorem 4.4.2. Since $\|f\| \leq |||f|||$ for all $f \in \mathcal{D}$ there exists a contraction $A : \mathcal{H}_1 \to \mathcal{H}$ such that $Af = f$ for all $f \in \mathcal{D}$. If $f \in \mathcal{H}_1$ and $Af = 0$, then there exist $f_n \in \mathcal{D}$ such that $|||f_n - f||| \to 0$ and $\|f_n\| \to 0$. Hence

$$\begin{aligned}\langle f, f\rangle_1 &= \lim_{m\to\infty}\lim_{n\to\infty}\langle f_m, f_n\rangle_1 \\ &= \lim_{m\to\infty}\lim_{n\to\infty}\{\langle f_m, f_n\rangle + \langle Hf_m, f_n\rangle\} \\ &= \lim_{m\to\infty}\{\langle f_m, 0\rangle\} \\ &= 0.\end{aligned}$$

Therefore A has zero kernel and $\mathcal{H}_1$ is embedded as a linear subspace of $\mathcal{H}$. The proof that Q is closable now follows from Theorem 4.4.2.

If K is the non-negative self-adjoint operator associated with the closure of Q we still have to prove that K is an extension of H. If $f, g \in \mathcal{D} \subseteq \mathrm{Dom}(K^{1/2})$, then

$$\langle K^{1/2}f, K^{1/2}g\rangle = \overline{Q'}(f,g) = \langle Hf, g\rangle.$$

The same holds by an approximation argument for all $f \in \mathcal{D}$ and $g \in \mathcal{H}_1$. We may now deduce that $\mathcal{D} \subseteq \mathrm{Dom}(K)$ and that $Hf = Kf$ for all $f \in \mathcal{D}$ by Lemma 4.4.1. □

Example 4.4.6 Let H be the symmetric operator defined on the subspace $\mathcal{D} := C_c^\infty(a,b)$ of $L^2(a,b)$ by $Hf = -f''$. H is not essentially self-adjoint but it has the associated quadratic form

$$Q(f) := \int_a^b |f'(x)|^2 \mathrm{d}x, \tag{4.4.4}$$

where $f \in \mathcal{D}$. We explain how to identify the Friedrichs extension K of H.

Let $\mathcal{E}$ be the space $C_0^\infty[a,b]$ of smooth functions on $[a,b]$ which vanish at a and b, and let $\tilde{Q}$ be the obvious extension of Q to $\mathcal{E}$. We claim that $\mathcal{E}$ is contained in the domain of the closure of Q and that $\overline{Q}$ equals $\tilde{Q}$ on $\mathcal{E}$. Let $k > 4/(b-a)$ and let $\{\phi_n\}_{n=k}^\infty$ be a sequence of functions in $\mathcal{D}$ such that $\phi_n(x) = 1$ if $a + 2/n < x < b - 2/n$, $\phi_n(x) = 0$ if $x < a + 1/n$, $\phi_n(x) = 0$ if $x > b - 1/n$, $0 \le \phi_n(x) \le 1$ for all x, and $|\phi_n'(x)| < cn$ for all x. Given $g \in \mathcal{E}$ define $g_n(x) = \phi_n(x)g(x)$, so that $g_n \in \mathcal{D}$. We have

$$\begin{aligned}\|g - g_n\|^2 &= \int_a^b |g|^2 |1 - \phi_n(x)|^2 \mathrm{d}x \\ &= o(1)\end{aligned}$$

as $n \to \infty$ by the dominated convergence theorem. Putting

$$A_n := \{x : a < x < a + 2/n \text{ or } b - 2/n < x < b\},$$

we have

$$\begin{aligned}\tilde{Q}(g - g_n) &= \int_a^b |(g - \phi_n g)'|^2 \mathrm{d}x \\ &\le 2\int_a^b |g'|^2 |1 - \phi_n|^2 \mathrm{d}x + 2\int_a^b |g|^2 |\phi_n'|^2 \mathrm{d}x \\ &\le 2\int_{A_n} |g'|^2 \mathrm{d}x + 2c^2 n^2 \int_{A_n} |g|^2 \mathrm{d}x \\ &= O(1/n)\end{aligned}$$

as $n \to \infty$ since $g(x) = 0$ and $|g'(x)| < \infty$ at $x = a, b$. This proves the claim at the start of the paragraph.

Now consider the set $\{f_n\}_{n=1}^\infty$ of Dirichlet eigenfunctions of H_D, defined in Example 1.2.3. These all lie in the space $\mathcal{E}$ and therefore lie in the domain of the closure $\overline{Q}$ of Q. It follows from Lemma 4.4.1 that f_n lie in the domain of the Friedrichs extension K and that $Kf_n = \{n^2\pi^2/(b-a)^2\}f_n$ for all $n \ge 1$. Thus K coincides with H_D on $\mathrm{lin}\{f_n : 1 \le n < \infty\}$. But both operators are essentially self-adjoint when restricted to this domain by Lemma 1.2.2, so $K = H_D$. □

Example 4.4.7 Let H_0 be the closure of $-\Delta$ initially defined on the dense linear subspace $C_c^\infty(\mathbf{R}^N)$ of $L^2(\mathbf{R}^N)$. Then the square root K of H_0, defined in Example 3.7.5, has domain $W^{1,2}(\mathbf{R}^N)$. If $f \in C_c^\infty$ then Fourier transforms and integration by parts yield

$$\langle H_0 f, f\rangle = \|Kf\|^2 = \int_{\mathbf{R}^N} |\nabla f|^2 \mathrm{d}^N x.$$

The quadratic form domain of H_0 is equal to $W^{1,2}(\mathbf{R}^N)$ and the quadratic form is given by

$$Q_0(f) = \int_{\mathbf{R}^N} |\nabla f|^2 \mathrm{d}^N x$$

for all f in that domain. □

4.5 The variational formulae

If one had to choose the single most computationally important result in spectral theory, it would without doubt be the variational formulae for eigenvalues, which originate from work of Rayleigh on elastic vibrations in 1873. These are the basis of fundamental numerical procedures for the computation of eigenvalues of $n \times n$ matrices and of the eigenvalues of complicated N-body molecular Hamiltonians in quantum chemistry. The emphasis in this book, however, will be on the use of the variational method to obtain quantitative estimates of eigenvalues, and for comparing the eigenvalues of different operators.

Let H be a non-negative self-adjoint operator on a Hilbert space $\mathcal{H}$ and let L be any finite-dimensional subspace of the domain of H. We define

$$\lambda(L) := \sup\{\langle Hf, f\rangle : f \in L \text{ and } \|f\| = 1\}.$$

These L-dependent numbers are used to define a non-decreasing sequence of numbers λ_n according to the Rayleigh–Ritz variational formula:

$$\lambda_n := \inf\{\lambda(L) : L \subseteq \mathrm{Dom}(H) \text{ and } \dim(L) = n\}. \qquad (4.5.1)$$

We shall see that in many situations it is possible to obtain explicit quantitative upper and lower bounds on these numbers. Their significance is explained in the following series of theorems.

Theorem 4.5.1 *Let H be a non-negative self-adjoint operator with compact resolvent. Then the numbers λ_n satisfy $\lim_{n\to\infty}\{\lambda_n\} = \infty$, and coincide with the eigenvalues of H written in increasing order and repeated according to multiplicity.*

Proof By Corollary 4.2.3 there exists a complete orthonormal set of eigenvectors $\{f_n\}_{n=1}^{\infty}$ of H with corresponding eigenvalues μ_n written in increasing order and repeated according to multiplicity. If $M_n = \mathrm{lin}\{f_r : 1 \le r \le n\}$ and $f \in M_n$, then

$$\begin{aligned}\langle Hf, f\rangle &= \sum_{r=1}^{n} \mu_r |\langle f, f_r\rangle|^2 \\ &\le \mu_n \sum_{r=1}^{n} |\langle f, f_r\rangle|^2 \\ &= \mu_n \|f\|^2.\end{aligned}$$

It follows that $\lambda_n \le \mu_n$.

Conversely let L be any n-dimensional linear subspace of $\mathrm{Dom}(H)$ and let P be the orthogonal projection on $\mathscr{H}$ with range M_{n-1}, so that P is given explicitly by

$$Pf := \sum_{r=1}^{n-1} \langle f, f_r\rangle f_r$$

for all $f \in \mathscr{H}$. The range of $P|_L$ has lower dimension than L so there must exist $f \in L$ with norm 1 such that $Pf = 0$. We then have $\langle f, f_r\rangle = 0$ for all $r \le n-1$. It follows that

$$\begin{aligned}\langle Hf, f\rangle &= \sum_{r=n}^{\infty} \mu_r |\langle f, f_r\rangle|^2 \\ &\ge \mu_n \sum_{r=n}^{\infty} |\langle f, f_r\rangle|^2 \\ &= \mu_n. \end{aligned} \tag{4.5.2}$$

We conclude that $\lambda(L) \ge \mu_n$ for all such subspaces, and hence that $\lambda_n \ge \mu_n$. □

If H is a self-adjoint $m \times m$ matrix, then λ_n are only defined for $n \le m$, but the above theorem is otherwise valid. In the infinite-dimensional case if we do not make the assumption that H has compact resolvent then the situation is more complicated.

Theorem 4.5.2 *Let H be a non-negative self-adjoint operator on $\mathscr{H}$, and let λ_m be defined by (4.5.1). If H has non-empty essential spectrum then one of the following cases occurs.*

(1) *There exists $a < \infty$ such that $\lambda_m < a$ for all m and $\lim_{m\to\infty}\{\lambda_m\} = a$. Then a is the smallest number in the essential spectrum, and the part of the spectrum of H in $[0, a)$ consists of the eigenvalues λ_m each repeated a number of times equal to its multiplicity.*

(2) *There exists $a < \infty$ and $N < \infty$ such that $\lambda_N < a$ but $\lambda_m = a$ for all $m > N$. Then a is the smallest number in the essential spectrum, and the part of the spectrum of H in $[0, a)$ consists of the eigenvalues $\lambda_1, ..., \lambda_N$ each repeated a number of times equal to its multiplicity.*

Proof If b is the smallest point in the essential spectrum of H then H has only isolated eigenvalues of finite multiplicity in $[0, b)$. Since their only possible limit point is b we may list them in increasing order, repeating each one according to its multiplicity. There are two cases to consider.

(1) If there are an infinite number of such eigenvalues, denoted $\{\mu_m\}_{m=1}^{\infty}$, then they converge monotonically to a limit in the essential spectrum, which must be b. The fact that $\lambda_m = \mu_m$ for all m follows as in Theorem 4.5.1, except for the proof of (4.5.2). The equation $Pf = 0$ implies that f has support in the set $\{(s, n) : h(s, n) \ge \mu_m\}$, following the notation of Theorem 2.5.1, and this implies (4.5.2). We finally observe that $a = b$.

(2) If there exist only a finite number M of such eigenvalues, then

$$0 \le \mu_1 \le \mu_2 \le ... \le \mu_M < b.$$

We see that $\lambda_m = \mu_m$ for all $m \le M$ and that $\lambda_m \ge b$ for $m > M$ by the same method. Since b lies in the essential spectrum of H, given $\varepsilon > 0$ the spectral subspace $L := L_{(b-\varepsilon, b+\varepsilon)}$ is infinite-dimensional, and there exists a countable orthonormal set $\{f_r\}_{r=1}^{\infty}$ in L. Applying the definition of λ_m to $\mathrm{lin}\{f_1, ..., f_m\}$, which is a subspace of $\mathrm{Dom}(H)$, we see that $\lambda_m \le b + \varepsilon$ for all m. Therefore $\lambda_m = b$ for all $m > M$. Finally we observe that $a = b$ and $M = N$. □

In all of the above theorems we assumed that the n-dimensional subspaces L are contained in the domain of H. We shall often use the variational formula to compare the eigenvalues of two different differential operators. This is best done at the quadratic form level, which involves considering subspaces L of $\mathrm{Dom}(H^{1/2})$. Let Q be a closed quadratic form and let $\mathscr{D}$ be a core for Q, that is a subspace of $\mathrm{Dom}(H^{1/2})$ which is dense in it for the norm of (5.4.2). If L is a finite-dimensional subspace of $\mathrm{Dom}(H^{1/2})$ then we define $\lambda(L)$ by

$$\lambda(L) := \sup\{Q(f) : f \in L \text{ and } \|f\| = 1\}.$$

The following theorem establishes the consistency of three different methods of defining λ_n.

Theorem 4.5.3 *If we put*

$$\lambda_n' := \inf\{\lambda(L) : L \subseteq \mathscr{D} \text{ and } \dim(L) = n\},$$
$$\lambda_n'' := \inf\{\lambda(L) : L \subseteq \mathrm{Dom}(H^{1/2}) \text{ and } \dim(L) = n\},$$

then $\lambda_n = \lambda_n' = \lambda_n''$ *for all* $n \geq 1$.

Proof Since $\mathscr{D} \subseteq \mathrm{Dom}(H^{1/2})$ we have $\lambda_n' \geq \lambda_n''$ for all $n \geq 1$. Conversely given $\varepsilon > 0$ let L be an n-dimensional subspace of $\mathrm{Dom}(H^{1/2})$ such that $\lambda(L) < \lambda_n'' + \varepsilon$. By restricting the form Q to $L \times L$ we see that there exists an orthonormal basis $f_1, \ldots, f_n$ of L such that $Q(f_i, f_j) = \gamma_i \delta_{i,j}$ for all $1 \leq i, j \leq n$ where $0 \leq \gamma_1 \leq \ldots \leq \gamma_n = \lambda(L)$. Since $\mathscr{D}$ is a core there exist $g_1, \ldots, g_n \in \mathscr{D}$ such that $|||f_i - g_i||| < \varepsilon$ for all $1 \leq i \leq n$. If $a_{i,j} := \langle g_i, g_j \rangle$ and $b_{i,j} := Q(g_i, g_j)$, then $|a_{i,j} - \delta_{i,j}| \leq c_n \varepsilon$ and $|b_{i,j} - \gamma_i \delta_{i,j}| \leq c_n' \varepsilon$. A direct matrix computation now shows that if $L' = \mathrm{lin}\{g_1, \ldots, g_n\}$, then

$$\begin{aligned}\lambda(L') &= \sup\left\{\sum_{i,j=1}^{n} b_{i,j}\alpha_i\bar{\alpha}_j : \sum_{i,j=1}^{n} a_{i,j}\alpha_i\bar{\alpha}_j \leq 1\right\} \\ &\leq \gamma_n + c_n''\varepsilon.\end{aligned}$$

Therefore $\lambda(L') \leq \lambda(L) + c_n''\varepsilon$. The proof of the equality $\lambda_n' = \lambda_n''$ is completed by taking the limit $\varepsilon \to 0$.

We claim that the equality $\lambda_n = \lambda_n''$ is a special case of the above result, since $\mathrm{Dom}(H)$ is a core for Q. In terms of the notation of Theorem 2.5.1 this is just the statement that the space

$$\left\{f : \int_{S\times\mathbf{N}} (1 + h(s,n)^2)|f(s,n)|^2 \mathrm{d}\mu < \infty\right\}$$

is dense in the space of functions for which the norm

$$|||f|||^2 := \int_{S\times\mathbf{N}} (1 + h(s,n))|f(s,n)|^2 \mathrm{d}\mu$$

is finite, given that h is non-negative μ-almost everywhere. □

From the numerical point of view it is not possible to enumerate all finite-dimensional subspaces of $\mathrm{Dom}(H)$ or of $\mathrm{Dom}(H^{1/2})$, and one proceeds in a slightly different way. One first selects a single finite-dimensional subspace $\mathscr{D}$ with as much sensitivity to the problem in hand as possible. The size of the subspace may involve a compromise between cost and accuracy. The choice is by no means simple, and a large part of theoretical quantum chemistry might be described as laying down rules for the selection of $\mathscr{D}$ for molecular Hamiltonians. The goal is to choose the subspace $\mathscr{D}$ in such a way that the constants E_n defined below are very good approximations to the eigenvalues λ_n. We will discuss this difficult matter in Section 4.6.

Lemma 4.5.4 *Let $\{f_n\}_{n=1}^N$ be an orthonormal basis of $\mathscr{D} \subseteq \mathrm{Dom}(H^{1/2})$ and let $\{E_n\}_{n=1}^N$ be the eigenvalues of the matrix*

$$A_{i,j} := Q(f_i, f_j)$$

written in increasing order and repeated according to multiplicity. Then $\lambda_n \le E_n$ for all $n \le N$.

Proof The eigenvalues $\{\lambda_n\}$ are determined by (4.5.1), while the numbers $\{E_n\}$ are determined by

$$E_n := \inf\{\lambda(L) : L \subseteq \mathscr{D} \text{ and } \dim(L) = n\}.$$

The lemma therefore follows from the inclusion $\mathscr{D} \subseteq \mathrm{Dom}(H^{1/2})$. □

The variational method may also be applied to the numerical computation of the eigenvalues of a non-negative self-adjoint $n \times n$ matrix A, considered as acting on $\mathbf{C}^n$ with the standard inner product. The smallest eigenvalue λ_1 is given by the formula

$$\lambda_1 = \min\{\langle A\phi, \phi\rangle : \|\phi\| = 1\}, \tag{4.5.3}$$

the corresponding eigenvector being the sequence $\phi \in \mathbf{C}^n$ which minimises the above expression. The formula (4.5.1) is not practical for computing higher eigenvalues. The variational method enables one to compute λ_m given the smallest $m-1$ eigenvalues $\lambda_1, ..., \lambda_{m-1}$ and the corresponding normalised eigenvectors $\phi_1, ..., \phi_{m-1}$. It is an immediate consequence of the spectral theorem that

$$\lambda_m = \inf\{\langle A\phi, \phi\rangle : \|\phi\| = 1 \text{ and } \langle\phi, \phi_r\rangle = 0 \text{ for } 1 \le r \le m-1\}. \quad (4.5.4)$$

Exercise 4.3 gives an alternative method of computing λ_m and is also proved using the spectral theorem.

4.6 Lower bounds on eigenvalues

Although the Rayleigh–Ritz formula is very useful for obtaining upper bounds on eigenvalues, it has definite limitations. These can be illustrated by reference to an example in one dimension which is close to being exactly soluble.

Example 4.6.1 Let $Hf(x) := -(\mathrm{d}/\mathrm{d}x)\{a(x)(\mathrm{d}f/\mathrm{d}x)\}$ on $L^2(0,1)$ subject to Dirichlet boundary conditions, where $c^{-1} \le a(x) \le c$ for all $x \in (0,1)$. The associated quadratic form Q satisfies the inequalities

$$c^{-1}\int_0^1 |f'(x)|^2\mathrm{d}x \le Q(f) := \int_0^1 a(x)|f'(x)|^2\mathrm{d}x \le c\int_0^1 |f'(x)|^2\mathrm{d}x$$

for all f in its domain $W_0^{1,2}(0,1)$, defined in Example 4.4.4. It follows that H has discrete spectrum and that its eigenvalues $\{\lambda_n\}_{n=1}^{\infty}$ satisfy

$$c^{-1}\pi^2 n^2 \le \lambda_n \le c\pi^2 n^2$$

for all $n \ge 1$. Our main point is that even if the coefficient $a(x)$ is a smooth function of x on $[0,1]$, the Rayleigh–Ritz formula does not allow one to compute the eigenvalues λ_n approximately without further information about the rate of variation of $a(x)$ as x increases. We consider henceforth only the first eigenvalue λ_1, which is given by the formula

$$\lambda_1 = \inf\left\{\int_0^1 a(x)|f'(x)|^2\mathrm{d}x : f \in B\right\},$$

where B is the set of $f \in W_0^{1,2}(0,1)$ such that $\|f\|_2 = 1$.
We now suppose that $a(x)$ is smooth and periodic on $\mathbf{R}$ with period 1, and define $a_n(x) := a(nx)$ for all $x \in (0,1)$. Let H_n be the corresponding differential operators on $L^2(0,1)$, all subject to Dirichlet boundary conditions. A similar argument to that of the Riemann–Lebesgue lemma shows that

$$\lim_{n\to\infty}\int_0^1 a_n(x)|f'(x)|^2\mathrm{d}x = \gamma\int_0^1 |f'(x)|^2\mathrm{d}x$$

for all $f \in W_0^{1,2}(0,1)$, where

$$\gamma := \int_0^1 a(s)\mathrm{d}s.$$

Therefore if we were to try to evaluate the first non-zero eigenvalue $\lambda_1(n)$ of H_n by selecting f from some fixed n-independent finite-dimensional subspace of $W_0^{1,2}(0,1)$, we would conclude that $\lim_{n\to\infty}\lambda_1(n) = \gamma\pi^2$. This formula is incorrect. In fact the convergence of the quadratic forms for all $f \in W_0^{1,2}(0,1)$ does not imply that H_n converges to $H_\infty := -\gamma\mathrm{d}^2/\mathrm{d}x^2$ in the norm resolvent or any other sense. The operators H_n in this example do converge, but to a quite different limit! We comment before continuing that this type of peculiar behaviour cannot occur if the coefficients converge almost everywhere; see Exercise 4.8. □

Theorem 4.6.2 *The operators H_n of the above example converge in the norm resolvent sense to $Hf := -\beta\mathrm{d}^2 f/\mathrm{d}x^2$ where*

$$\beta^{-1} := \int_0^1 a(s)^{-1}\mathrm{d}s$$

is a constant satisfying $c^{-1} \le \beta < \gamma \le c$ unless a is a constant function.

Proof The operator H_n is invertible and its inverse is given by

$$H_n^{-1}f(x) = \int_0^1 G_n(x,y)f(y)\mathrm{d}y$$

where the Green function $G_n(x,y)$ has the following standard form, proved in any course on Sturm–Liouville operators.

$$G_n(x,y) := \begin{cases} k_n\phi_n(x)\psi_n(y) & \text{if } x \le y, \\ k_n\psi_n(x)\phi_n(y) & \text{if } x \ge y, \end{cases}$$

where

$$\phi_n(x) := \int_0^x a_n(s)^{-1}\mathrm{d}s,$$

$$\psi_n(x) := \int_x^1 a_n(s)^{-1}\mathrm{d}s,$$

$$k_n^{-1} := \int_0^1 a_n(s)^{-1}\mathrm{d}s.$$

We observe that $G_n(x,y) = 0$ if $x = 0$ or $y = 1$.

Now let us define the limiting Green function, namely

$$G(x, y) := \begin{cases} k\phi(x)\psi(y) & \text{if } x \leq y, \\ k\psi(x)\phi(y) & \text{if } x \geq y, \end{cases}$$

where

$$\begin{aligned} \phi(x) &:= \beta^{-1}x, \\ \psi(x) &:= \beta^{-1}(1-x), \\ k &:= \beta. \end{aligned}$$

It is straightforward to demonstrate that ϕ_n converges uniformly to ϕ, that ψ_n converges uniformly to ψ and that k_n converges to k as $n \to \infty$. It follows that G_n converges uniformly to G on $(0,1) \times (0,1)$. Since G is the Green function of H, we see that

$$\lim_{n\to\infty} \|H_n^{-1} - H^{-1}\| = 0.$$

This is enough to prove norm resolvent convergence; see Lemma 2.6.1 and Exercise 2.10. □

We treated the above example in detail in order to point out the limitations of the Rayleigh–Ritz formula. It provides upper bounds to the eigenvalues of second order elliptic operators, but they need to be supplemented by lower bounds or error estimates. One method of obtaining lower bounds, due to Temple and refined by Lehmann, depends upon the use of test functions from the operator domain rather than just from the quadratic form domain. It also depends upon having crude information about the location of higher eigenvalues.

Let H be a non-negative self-adjoint operator on a Hilbert space $\mathcal{H}$. Suppose that H has eigenvalues $\{\lambda_n\}_{n=1}^{\infty}$ written in increasing order and repeated according to multiplicity. Suppose that ϕ is a function which is, in fact, a good approximation to the eigenfunction associated with the eigenvalue λ_n, and suppose that we wish to use it to obtain a lower bound on λ_n. Our theorem requires a crude lower bound ρ to the next eigenvalue λ_{n+1}.

Theorem 4.6.3 *Let* $\phi \in \mathrm{Dom}(H)$ *and suppose that*

$$\lambda_n \leq \frac{\langle H\phi, \phi\rangle}{\langle \phi, \phi\rangle} < \rho \leq \lambda_{n+1}.$$

Then

$$\lambda_n \geq \frac{\rho\langle H\phi, \phi\rangle - \langle H\phi, H\phi\rangle}{\rho\langle \phi, \phi\rangle - \langle H\phi, \phi\rangle}$$

with equality if ϕ is the eigenfunction of H associated with the eigenvalue λ_n.

Proof Since $\mathrm{Spec}(H)\cap(\lambda_n, \rho) = \emptyset$ the self-adjoint operator $(H-\rho)(H-\lambda_n)$ is non-negative. Therefore

$$\begin{aligned} 0 &\leq \langle (H-\rho)\phi, (H-\lambda_n)\phi\rangle \\ &= \langle H\phi, H\phi\rangle - \rho\langle H\phi, \phi\rangle - \lambda_n\langle H\phi, \phi\rangle + \rho\lambda_n\langle\phi,\phi\rangle. \end{aligned}$$

The statement of the theorem is a simple reorganisation of this inequality. □

The implementation of the Temple–Lehmann lower bounds for particular partial differential operators is necessarily rather delicate, since the goal is to obtain accurate upper and lower bounds at as little computational cost as possible. If all of the calculations are performed using interval arithmetic, one may even obtain rigorous upper and lower bounds on the eigenvalue. This is properly a matter for numerical analysis texts rather than this volume.

While we are discussing limitations of the Rayleigh–Ritz formula, we mention an even more serious one. The Schrödinger operator which describes a periodic crystal with a single localised impurity may have a spectrum of the type

$$[a, b] \cup \{\mu_1, ..., \mu_k\} \cup [c, \infty),$$

where $a < b < \mu_1 < ... < \mu_k < c$. The intervals $[a, b]$ and $[c, \infty)$, called bands, are due to the periodic structure, while the isolated eigenvalues μ_i of finite multiplicity are a result of the impurity. It is quite impossible to obtain any information about μ_i using the Rayleigh–Ritz formula since Theorem 4.5.2 yields $\lambda_m = a$ for all m. The problem of determining the location of bound states in the gaps between bands is a very difficult one, and the subject of much current research.

Exercises

4.1 Let $E_1, ..., E_n$ be a sequence of disjoint Borel sets in $\mathbf{R}^N$ of positive finite measure, and let $\chi_1, ..., \chi_n$ be their characteristic functions. Given $1 \leq p < \infty$ prove that the operator P on $L^p(\mathbf{R}^N)$ defined by

$$Pf := \sum_{r=1}^{n} |E_r|^{-1}\langle f, \chi_r\rangle\chi_r$$

is a projection of finite rank, and find its norm and range. Use a sequence of such projections to extend Lemma 4.2.1 to $L^p(\mathbf{R}^N)$.

4.2 Let H be a non-negative self-adjoint operator on the Hilbert space $\mathcal{H}$ with form domain $\mathcal{Q} := \mathrm{Dom}(H^{1/2})$. Prove that H has compact resolvent if and only if the canonical embedding $i : \mathcal{Q} \to \mathcal{H}$ is a compact linear operator. Prove also that H has compact resolvent if and only if $H^{1/2}$ has compact resolvent.

4.3 Let $\lambda_1, ..., \lambda_n$ be the eigenvalues of an $n \times n$ self-adjoint matrix A, written in increasing order as usual. Show that for any $m \le n$ one has

$$\sum_{r=1}^{m} \lambda_r = \min\{\mathrm{tr}(L) : \dim(L) = m\},$$

where L denotes any linear subspace of $\mathbf{C}^n$, and

$$\mathrm{tr}(L) := \sum_{r=1}^{m} Q(\phi_r)$$

for some (any) orthonormal basis $\{\phi_r\}_{r=1}^m$ of L.

4.4 Use Lemma 4.2.1 to prove the following statements.

(a) The set $\mathcal{I}$ of compact operators on a Hilbert space $\mathcal{H}$ is a norm closed linear subspace of the space $\mathcal{L}(\mathcal{H})$ of all bounded linear operators on $\mathcal{H}$.

(b) If $A \in \mathcal{I}$ and $B \in \mathcal{L}(\mathcal{H})$, then $AB \in \mathcal{I}$ and $BA \in \mathcal{I}$.

(c) If $A \in \mathcal{I}$, then $A^* \in \mathcal{I}$.

4.5 Let $A : L^2(\mathbf{R}^N) \to L^2(\mathbf{R}^N)$ be a Hilbert–Schmidt operator, that is be of the form

$$Af(x) := \int_{\mathbf{R}^N} a(x,y) f(y) \mathrm{d}^N y,$$

where $a \in L^2(\mathbf{R}^N \times \mathbf{R}^N)$. By expanding a in the form

$$a(x,y) = \sum_{m,n=1}^{\infty} c_{m,n} \phi_m(x) \phi_n(y),$$

where $\{\phi_n\}_{n=1}^{\infty}$ is a complete orthonormal set in $L^2(\mathbf{R}^N)$, and using Lemma 4.2.1, prove that A is compact, and that $\|A\| \le \|A\|_{\mathrm{HS}}$, where $\|A\|_{\mathrm{HS}}^2 = \int_{R^N \times R^N} |a(x,y)|^2 \mathrm{d}^N x \mathrm{d}^N y$.

4.6 Let H be a non-negative self-adjoint operator on $\mathcal{H}$ whose form Q has domain $\mathcal{D}$. Let $A \ge 0$ be a bounded self-adjoint operator on $\mathcal{H}$ and define $\tilde{Q}$ on $\mathcal{D}$ by

$$\tilde{Q}(f) := Q(f) + \langle Af, f \rangle.$$

Prove that $\tilde{Q}$ is closed on $\mathcal{D}$. Use Lemma 4.4.1 to prove that the associated self-adjoint operator $\tilde{H}$ has domain equal to the domain of H and that $\tilde{H}$ coincides with the operator $(H + A)$ defined in Theorem 1.4.2.

4.7 Let H_n be defined on $L^2(0,1)$ as in Example 4.6.1 with

$$a_n(x) := 2 + \sin(2n\pi x).$$

Find the operator H to which H_n converges in the norm resolvent sense. If λ_1 is the smallest eigenvalue of H and ϕ is the corresponding normalised eigenfunction, prove by a direct computation that $\langle H_n\phi, \phi \rangle$ does not converge to λ_1 as $n \to \infty$.

4.8 Let H_n be defined on $L^2(a,b)$ for $n \geq 0$ by

$$H_n f(x) := -\frac{\mathrm{d}}{\mathrm{d}x}\left\{a_n(x)\frac{\mathrm{d}f}{\mathrm{d}x}\right\}$$

subject to Dirichlet boundary conditions, where $0 < c^{-1} \leq a_n(x) \leq c < \infty$ for all $x \in (a,b)$ and all n. Suppose that $\lim_{n\to\infty} a_n(x) = a_0(x)$ almost everywhere in (a,b). By considering the behaviour of the Green functions as in Theorem 4.6.2, prove that H_n converge to H_0 in the norm resolvent sense.

4.9 Let H be a non-negative self-adjoint operator on a Hilbert space $\mathcal{H}$, and let $\lambda > 0$. Use the resolvent equations to prove that $(H + \lambda)^{-1}$ is compact if and only if $(H + 1)^{-1}$ is compact.

5
Further spectral results

5.1 The Poisson problem

This chapter is devoted to a miscellany of spectral problems, which are applications of the ideas of Chapters 2 and 4. The first of these is called the Poisson problem. The Poisson and Dirichlet problems both date from the first half of the nineteenth century, but decisive developments in their solution have taken place in this century.

Let Ω be a bounded region in $\mathbf{R}^N$ with boundary $\partial\Omega$. Let f be a continuous function on $\overline{\Omega}$ and let g be a continuous function on $\partial\Omega$. Poisson's problem is then the determination of a C^2 function h on $\overline{\Omega}$ such that $h|_{\partial\Omega} = g$ and $-\Delta h = f$ on Ω; the case $f = 0$ is called the Dirichlet problem. The existence of a solution is a very delicate issue for general regions Ω, and is properly a part of potential theory rather than spectral theory.

There are at least three different sources of such problems in mathematical physics. One can look for the electrostatic potential h on Ω associated with a charge distribution f inside Ω. One can try to find the equilibrium temperature distribution in a piece of matter subjected to external heating. Finally one can ask about the displacement of an elastic membrane subject to small transverse forces.

We shall not try to give a thorough treatment of the Poisson problem, but will show how it is related to issues of spectral theory, under reasonable conditions on the functions and region involved. Suppose first that g is the restriction of some C^2 function k on $\overline{\Omega}$ and put $h_0 := h - k$ and $f_0 := f + \Delta k$. A straightforward computation shows that the Poisson problem is equivalent to the existence of a C^2 function h_0 on $\overline{\Omega}$ such that $h_0|_{\partial\Omega} = 0$ and $-\Delta h_0 = f_0$ in Ω. The formal solution of this problem is $h_0 := H^{-1} f_0$, where $H = -\Delta$ in $L^2(\Omega)$ subject to Dirichlet boundary conditions. Regularity properties of h inside Ω depend upon obtaining appropriate information about the operator H^{-1}.

The remainder of this section is devoted to comments relevant to the approximate solution of the equation $Hh_0 = f_0$; we drop the subscript 0 for simplicity. In numerical analysis the method we shall describe is called the Ritz–Galerkin method.

Lemma 5.1.1 *Let H be a self-adjoint operator on a real Hilbert space $\mathcal{H}$ and suppose that $H \geq c1$ for some $c > 0$. Given $f \in \mathcal{H}$ the solution of $Hh = f$ is equal to the point at which the function $F : \mathrm{Dom}(H) \to \mathbf{R}$ defined by*

$$F(\phi) := \langle H\phi, \phi\rangle - 2\langle \phi, f\rangle$$

takes its minimum value.

Proof The assumption on H implies that $0 \notin \mathrm{Spec}(H)$, so that $h := H^{-1}f$ exists and is in $\mathrm{Dom}(H)$. We may then rewrite F in the form

$$F(\phi) = \langle H(\phi - h), \phi - h\rangle - \langle Hh, h\rangle$$

which yields the statement of the lemma immediately. □

One of the ways of obtaining approximate solutions to $Hh = f$ is to look for the best approximation to h within a specified finite-dimensional subspace L. This subspace may consist of continuous piecewise linear functions constructed by the finite element method, or any other class of functions considered to be appropriate to the problem. We rewrite F in the form $F(\phi) := Q(\phi) - 2\langle \phi, f\rangle$, and take L to be any finite-dimensional subspace of $\mathrm{Dom}(H^{1/2})$.

Lemma 5.1.2 *Under the above assumptions, the restriction of F to L is convex and has a unique minimum, at $h_0 \in L$. The element h_0 is the closest point in L to h in the sense that it minimises the function $\phi \in L \to Q(h - \phi)$.*

Proof Let $\{e_i\}_{i=1}^n$ be a complete orthonormal set in L. Every $\phi \in L$ may be expanded in the form $\phi = \sum_{i=1}^n \alpha_i e_i$ and we then have

$$F(\phi) = \sum_{i,j=1}^{n} A_{i,j}\alpha_i\alpha_j - 2\sum_{i=1}^{n} \alpha_i f_i,$$

where $\{A_{i,j}\}$ is a real symmetric matrix all of whose eigenvalues are greater than or equal to $c > 0$. It follows from the spectral theorem that

F is strictly convex and possesses a unique minimum. Differentiating F at the point h_0 where it minimises, we obtain $Q(h_0, \phi) - \langle f, \phi \rangle = 0$ for all $\phi \in L$. Subtracting the similar equation $Q(h, \phi) - \langle f, \phi \rangle = 0$ valid for all $\phi \in \mathrm{Dom}(H^{1/2})$, we obtain $Q(h - h_0, \phi) = 0$ for all $\phi \in L$. The proof is now a consequence of standard geometrical arguments in the space $\mathrm{Dom}(H^{1/2})$ provided with the inner product Q. □

5.2 The heat equation

If H is a non-negative self-adjoint operator on a Hilbert space $\mathscr{H}$, then the abstract heat equation

$$\frac{\partial f}{\partial t} = -H\{f(t)\}$$

subject to the initial condition $f(0) = a$, has the formal solution

$$f(t) = \mathrm{e}^{-Ht} a.$$

We discuss the precise interpretation of the heat equation in Exercises 5.3 and 5.4. We can, however, define $T_t := \mathrm{e}^{-Ht}$ for $t \geq 0$ by using the functional calculus of Chapter 2. The point is that for all $t \geq 0$ the function $f_t : \mathbf{R} \to \mathbf{R}$ defined by

$$f_t(u) := \begin{cases} \mathrm{e}^{-tu} & \text{if } u \geq 0 \\ 1 & \text{otherwise} \end{cases}$$

is bounded and continuous. Since $\mathrm{Spec}(H) \subseteq [0, \infty)$, we could equally well take any other bounded definition of $f_t(u)$ when $u < 0$.

Theorem 5.2.1 *The operator T_t is a self-adjoint contraction on $\mathscr{H}$ for all $t \geq 0$. If $0 < s < t < \infty$, then*

$$\|T_t - T_s\| \leq \frac{t-s}{\mathrm{e}s}, \tag{5.2.1}$$

so $t \to T_t$ is norm continuous for $t > 0$. Also

$$\lim_{t \to 0} \|T_t f - f\| = 0 \tag{5.2.2}$$

for all $f \in \mathscr{H}$. The operators $\{T_t\}_{t \geq 0}$ form a semigroup in the sense that

$$T_{s+t} = T_s T_t$$

for all $s, t \geq 0$.

Proof Since $0 \le f_t(u) \le 1$ for all $u \in \mathbf{R}$ the first statement follows from Theorem 2.5.3. If we define $g := f_s - f_t$ then $g(u) = 0$ for all $u \le 0$. If $u > 0$ then

$$|g(u)| = \mathrm{e}^{-su}\left(1 - \mathrm{e}^{(s-t)u}\right) \le (t-s)u\mathrm{e}^{-su} \le \frac{t-s}{\mathrm{e}s}.$$

The bound (5.2.1) now follows by applying Theorem 2.5.3.

It follows from Theorems 2.5.1 and 2.5.3 that

$$\|T_t f - f\|^2 = \int_{S\times\mathbf{N}} \left|Uf(z)\left(1 - \mathrm{e}^{-h(z)t}\right)\right|^2 \mu(\mathrm{d}z),$$

where U is unitary and $h(z) \ge 0$ for all $z \in S \times \mathbf{N}$. The fact that this converges to 0 as $t \to 0$ is proved using the dominated convergence theorem. The semigroup property of the family $\{T_t\}_{t\ge0}$ follows immediately from Theorem 2.5.3. □

One of the important questions about the heat equation concerns the rate of convergence of solutions to equilibrium as $t \to \infty$. This can be investigated by means of the spectral theorem.

Theorem 5.2.2 *Let P_0 be the orthogonal projection of $\mathcal{H}$ onto the closed subspace $\mathcal{L} := \{f \in \mathrm{Dom}(H) : Hf = 0\}$. Then*

$$\lim_{t\to\infty} \|T_t f - P_0 f\| = 0$$

for all $f \in \mathcal{H}$. If $\lambda > 0$ and $P_{[\lambda,\infty)}f = f$, then $P_0 f = 0$ and

$$\|T_t f\| \le \mathrm{e}^{-\lambda t}\|f\|$$

for all $t \ge 0$. If $\mathrm{Spec}(H) \subseteq [\lambda,\infty)$, then $\|T_t\| \le \mathrm{e}^{-\lambda t}$ for all $t \ge 0$.

These statements are all immediate applications of the spectral theorem. They provide motivation for investigating when 0 does not lie in the spectrum of a non-negative self-adjoint operator, and when 0 is an isolated point of the spectrum. Both of these questions will be studied in later sections.

If 0 lies in the spectrum of H but is not an eigenvalue, then $\lim_{t\to\infty} \|T_t f\| = 0$ for all $f \in \mathcal{H}$, but it is impossible to give any useful bounds on the rate of convergence to 0 valid for all $f \in \mathcal{H}$. To achieve this we go outside the purely Hilbert space context.

Example 5.2.3 Let $H := -\overline{\Delta}$ acting in $L^2(\mathbf{R}^N)$. Then the analysis of Chapter 3 (particularly Theorem 3.5.5) can be used to show that $T_t f = k_t * f$ for all $t > 0$ where

$$k_t(x) := (4\pi t)^{-N/2} \mathrm{e}^{-|x|^2/4t}$$

and

$$(2\pi)^{N/2} \hat{k}_t(y) = \mathrm{e}^{-t|y|^2}.$$

This formula may be used to control the rate of convergence of $\|T_t f\|$ to 0 for a dense set of $f \in L^2$. □

Theorem 5.2.4 *In the above example we have*

$$\|T_t f\|_2 \le c t^{-N/4} \|f\|_1$$

for all $f \in L^1 \cap L^2$ and all $t > 0$.

Proof We have

$$\begin{aligned} \|T_t f\|_2 &= \|\mathscr{F} T_t f\|_2 \\ &= \|(2\pi)^{N/2} \hat{k}_t (\mathscr{F} f)\|_2 \\ &\le c_1 \|\hat{k}_t\|_2 \|\mathscr{F} f\|_\infty \\ &\le c_1 \|\hat{k}_t\|_2 \|f\|_1. \end{aligned}$$

The proof is completed with the computation

$$\begin{aligned} \|\hat{k}_t\|_2^2 &= c_2 \int_0^\infty \mathrm{e}^{-2r^2 t} r^{N-1} \mathrm{d}r \\ &= c_3 \int_0^\infty \mathrm{e}^{-2s^2} t^{-N/2} s^{N-1} \mathrm{d}s \\ &= c_4 t^{-N/2}. \end{aligned}$$

□

The property proved in the above theorem is a particular case of the notion of ultracontractivity, which is investigated in depth in Davies (1989), and has proved of great importance for clarifying the spectral behaviour of more general second order elliptic operators. Another way of looking at the long time behaviour of Example 5.2.3 is given in the next theorem. It also is the subject of current research for more general operators.

Theorem 5.2.5 *Let $0 \le f \in L^1 \cap L^2$ not be identically zero. Then there exists a constant $c > 0$ such that*

$$c^{-1} t^{-N/2} \le \langle T_t f, f \rangle \le c t^{-N/2}$$

for all $t \ge 1$.

Proof We have

$$\begin{aligned}
\langle T_t f, f \rangle &= \int_{\mathbf{R}^{2N}} k_t(x-y) f(y) f(x) \, \mathrm{d}^N x \, \mathrm{d}^N y \\
&\le (4\pi t)^{-N/2} \int_{\mathbf{R}^{2N}} f(y) f(x) \, \mathrm{d}^N x \, \mathrm{d}^N y \\
&= (4\pi t)^{-N/2} \|f\|_1^2
\end{aligned}$$

for all $t > 0$. Conversely let A be a bounded measurable set with positive measure $|A|$ such that $f(x) \ge a > 0$ for some $a > 0$ and all $x \in A$. Define $b > 0$ by

$$b := \inf\{ \mathrm{e}^{-(x-y)^2/4} : x, y \in A \}.$$

Then

$$\begin{aligned}
\langle T_t f, f \rangle &\ge a^2 \int_{A \times A} (4\pi t)^{-N/2} \mathrm{e}^{-(x-y)^2/4t} \, \mathrm{d}^N x \, \mathrm{d}^N y \\
&\ge (4\pi t)^{-N/2} a^2 b |A|^2
\end{aligned}$$

for all $t \ge 1$. □

5.3 The Hardy inequality

Although the original Hardy inequality was stated and proved only for functions on a half-line, it has given rise to many generalizations and applications to partial differential operators. There is even an entire book by Opic and Kufner (1990) devoted to the study of its various forms. Our goal here is to give some simple versions of the inequality and some indications of their usefulness. The case $\alpha = 0$ of the lemma below is the classical version of the inequality of Hardy (1920). The case $\alpha = 1$ is rather special and is treated in Exercise 5.6.

Lemma 5.3.1 *Let $0 < b < \infty$ and let f be a smooth function on $[0, b]$ which vanishes in some neighbourhood of 0. Then*

$$\frac{(1-\alpha)^2}{4} \int_0^b x^{\alpha-2} |f(x)|^2 \mathrm{d}x \le \int_0^b x^\alpha |f'(x)|^2 \mathrm{d}x,$$

provided $-\infty < \alpha < 1$. A similar bound holds for all $\alpha \in \mathbf{R}$ and $0 < b \le \infty$ if $f \in C_c^\infty(0,b)$.

Proof We assume that f is real-valued, the general case being deducible from this. Put $\lambda := (1-\alpha)/2 > 0$. Then

$$\begin{aligned}\int_0^b x^\alpha |f'|^2 \mathrm{d}x &= \int_0^b x^\alpha \left| x^\lambda (x^{-\lambda} f)' + \lambda x^{-1} f \right|^2 \mathrm{d}x \\ &\ge \lambda^2 \int_0^b x^{\alpha-2} |f|^2 \mathrm{d}x + 2\lambda \int_0^b (x^{-\lambda} f)' x^{\alpha-1+\lambda} f \mathrm{d}x \\ &= \lambda^2 \int_0^b x^{\alpha-2} |f|^2 \mathrm{d}x + \lambda \left[(x^{-\lambda} f)^2 \right]_0^b \\ &\ge \lambda^2 \int_0^b x^{\alpha-2} |f|^2 \mathrm{d}x. \end{aligned}$$ □

The condition $\lambda > 0$ was used to control the indefinite integral at $x = b$, and is not necessary if f vanishes in a neighbourhood of b.

Corollary 5.3.2 *Let $a < b$ and put $d(x) := \min\{x - a, b - x\}$ for all $a < x < b$. Then*

$$\int_a^b \frac{|f(x)|^2}{4d(x)^2} \mathrm{d}x \le \int_a^b |f'(x)|^2 \mathrm{d}x.$$

for all $f \in C_c^\infty(a,b)$.

Proof We apply the previous lemma to the intervals $[a,(a+b)/2]$ and $[(a+b)/2, b]$ after reversing the direction of the latter interval. □

Lemma 5.3.3 *Let $0 < b < \infty$ and let f be a smooth function on $[0,b]$ which vanishes in some neighbourhood of 0. Then*

$$\frac{9}{16} \int_0^b x^{-4} |f(x)|^2 \mathrm{d}x \le \int_0^b |f''(x)|^2 \mathrm{d}x.$$

A similar bound holds if $b = \infty$ and $f \in C_c^\infty(0,\infty)$.

Proof We combine Lemma 5.3.1 applied to f and $\alpha = -2$ with the same lemma applied to f' and $\alpha = 0$. □

The most important application of the Hardy inequality is as a technical tool in more advanced theoretical studies of elliptic operators. However, it can also be used directly to determine whether or not 0 lies in the spectrum of a non-negative self-adjoint operator. We start by considering a region with a simple enough shape that bounds of this type can be obtained without using the Hardy inequality.

Theorem 5.3.4 *Let $f : \mathbf{R}^N \to [0,\infty)$ be a continuous function and let*

$$\Omega := \{(x,y) \in \mathbf{R}^N \times \mathbf{R}^M : |y| < f(x)\}.$$

Let H be the Friedrichs extension of $-\Delta$ initially defined on $C_c^\infty(\Omega) \subseteq L^2(\Omega)$. Then

$$\frac{M\pi^2}{4} \int_\Omega \frac{|\phi(x,y)|^2}{f(x)^2} \mathrm{d}^N x \, \mathrm{d}^M y \le \int_\Omega |\nabla \phi|^2 \mathrm{d}^N x \, \mathrm{d}^M y \tag{5.3.1}$$

for all $\phi \in C_c^\infty(\Omega)$. If f is bounded, then $\mathrm{Spec}(H) \subseteq [\lambda, \infty)$ *where*

$$\lambda := \frac{M\pi^2}{4\|f\|_\infty^2} > 0.$$

Proof Put $y' := (y_2, y_3, ..., y_M)$ so that $y = (y_1, y')$. The interval

$$I(x,y') := \{y_1 \in \mathbf{R} : (x,y) \in \Omega\}$$

has length at most $2f(x)$. A combination of Example 1.2.3 and Theorem 4.3.1 applied to $-\partial^2/\partial y_1^2$ on this interval yields the inequality

$$\frac{\pi^2}{4} \int_{I(x,y')} \frac{|\phi(x,y)|^2}{f(x)^2} \mathrm{d}y_1 \le \int_{I(x,y')} \left| \frac{\partial \phi}{\partial y_1} \right|^2 \mathrm{d}y_1$$

for all $\phi \in C_c^\infty(\Omega)$. By integrating with respect to the remaining variables we obtain the bound

$$\frac{\pi^2}{4} \int_\Omega \frac{|\phi(x,y)|^2}{f(x)^2} \mathrm{d}^N x \, \mathrm{d}^M y \le \int_\Omega \left| \frac{\partial \phi}{\partial y_1} \right|^2 \mathrm{d}^N x \, \mathrm{d}^M y.$$

We now replace y_1 by y_i and sum over all values of i to obtain (5.3.1). The same bound now extends to all ϕ in the domain of the Friedrichs extension. If f is bounded we deduce that

$$\lambda \|\phi\|_2^2 \le \langle H\phi, \phi \rangle$$

for all $\phi \in \mathrm{Dom}(H)$, which suffices to prove the lower bound to the spectrum of H. □

If we consider regions with more complicated shapes, then the method just described may not be easy to adapt, and a multi-dimensional version of the Hardy inequality becomes important. We shall prove this only in two dimensions; an analogous treatment in higher dimensions is straightforward provided one is familiar with integration with respect to the Haar measure of the orthogonal group. The inequality is proved in two stages, the first of which does not impose any conditions on the region.

Theorem 5.3.5 *Let Ω be a region in $\mathbf{R}^2$ and let $f \in C_c^\infty(\Omega)$. Then*

$$\frac{1}{2}\int_\Omega \frac{|f(x)|^2}{m(x)^2}\mathrm{d}^2x \le \int_\Omega |\nabla f(x)|^2 \mathrm{d}^2x,$$

where the pseudodistance $m(x)$ is defined by

$$\frac{1}{m(x)^2} := \frac{1}{2\pi}\int_{-\pi}^{\pi} \frac{\mathrm{d}\theta}{d_\theta(x)^2}$$

and $d_\theta : \Omega \to (0, +\infty]$ is defined by

$$d_\theta(x) := \min\{|s| : x + s\mathrm{e}^{i\theta} \notin \Omega\}.$$

Proof Let ∂_θ denote partial differentiation in the direction $\mathrm{e}^{i\theta}$. An application of Corollary 5.3.2 yields

$$\frac{1}{4}\int_\Omega \frac{|f|^2}{d_\theta^2}\mathrm{d}^2x \le \int_\Omega |\partial_\theta f|^2 \mathrm{d}^2x$$

for all $f \in C_c^\infty(\Omega)$. Hence

$$\frac{1}{4}\int_\Omega \left(\frac{|f|^2}{d_\theta^2} + \frac{|f|^2}{d_{\theta+\pi/2}^2}\right)\mathrm{d}^2x \le \int_\Omega |\nabla f|^2 \mathrm{d}^2x.$$

The theorem follows by integrating this with respect to θ over $(-\pi, \pi)$. □

The above theorem only becomes useful upon obtaining some information connecting the pseudodistance $m(x)$ with the true distance

$$d(x) := \min\{|x - y| : y \notin \Omega\}.$$

We say that Ω is regular if there exists a constant $c < \infty$ such that

$$d(x) \le m(x) \le c d(x)$$

for all $x \in \Omega$, the first inequality being automatic. There are many different sufficient conditions for regularity, but the following one is particularly simple.

Theorem 5.3.6 *The region $\Omega \subseteq \mathbf{R}^2$ is regular if there exists a constant $c > 0$ such that*

$$\left|\{y \notin \Omega : |y - a| < r\}\right| \ge 2cr^2$$

for all $a \in \partial\Omega$ and all $r > 0$.

Proof Given $x \in \Omega$ let $r := d(x) = |x - a|$ for some $a \in \partial\Omega$. Let S be the set of $\theta \in (-\pi, \pi)$ such that

$$x + se^{i\theta} \in \{y \notin \Omega : |y - a| < r\}$$

for some $s > 0$. This condition implies $r < s < 2r$.

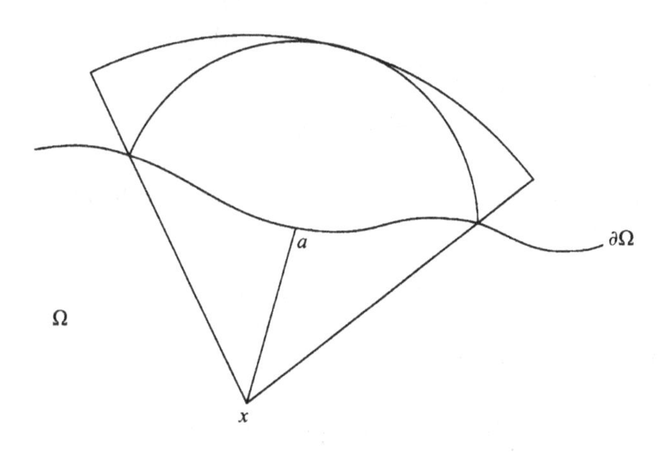

We have

$$2|S|r^2 \geq \left|\{y \notin \Omega : |y - a| < r\}\right| \geq 2cr^2$$

so $|S| \geq c$. We deduce that

$$\begin{aligned}\frac{1}{m(x)^2} &= \frac{1}{2\pi}\int_{-\pi}^{\pi} \frac{\mathrm{d}\theta}{d_\theta(x)^2} \\ &\geq \frac{1}{2\pi}\int_S \frac{\mathrm{d}\theta}{d_\theta(x)^2} \\ &\geq \frac{|S|}{8\pi d(x)^2}.\end{aligned}$$

This implies that

$$m(x) \leq \left(\frac{8\pi}{c}\right)^{1/2} d(x)$$

for all $x \in \Omega$. The lower bound on $m(x)$ is elementary. □

Theorem 5.3.7 *Let $\Omega \subseteq \mathbf{R}^2$ be regular, and let H be the Friedrichs extension of $-\Delta$ initially defined on $C_c^\infty(\Omega)$. Then $0 \in \mathrm{Spec}(H)$ if and only if the inradius of Ω is infinite.*

Proof Suppose that Ω is regular with constant c_0 and that it has finite inradius $r := \sup\{d(x) : x \in \Omega\}$. Then

$$\frac{1}{2c_0^2 r^2}\int_\Omega |f|^2 \mathrm{d}^2x \le \frac{1}{2}\int_\Omega \frac{|f|^2}{m^2}\mathrm{d}^2x \le \int_\Omega |\nabla f|^2 \mathrm{d}^2x$$

for all $f \in C_c^\infty(\Omega)$. This implies that

$$\mathrm{Spec}(H) \subseteq \left[\frac{1}{2c_0^2 r^2}, \infty\right).$$

Conversely suppose that $d(x)$ is an unbounded function. Let $\phi \in C_c^\infty(\mathbf{R}^2)$. For any $r > 0$ there exists $a \in \mathbf{R}^2$ such that the function ψ defined by $\psi(x) := \phi((x-a)/r)$ has support in Ω. Now

$$\frac{\int_\Omega |\nabla\psi|^2 \mathrm{d}^2x}{\int_\Omega |\psi|^2 \mathrm{d}^2x} = \frac{\int_{\mathbf{R}^2} |\nabla\phi|^2 \mathrm{d}^2x}{r^2 \int_{\mathbf{R}^2} |\phi|^2 \mathrm{d}^2x}$$

and this converges to 0 as $r \to \infty$. Therefore

$$\inf\left\{\int_\Omega |\nabla f|^2 \mathrm{d}^2x \Big/ \int_\Omega |f|^2 \mathrm{d}^2x : f \in C_c^\infty(\Omega)\right\} = 0$$

and $0 \in \mathrm{Spec}(H)$. □

5.4 Singular elliptic operators

Many important elliptic operators are singular elliptic, in the sense that their coefficients diverge to 0 or ∞ at the boundary of the region in question. We shall use the Hardy inequality to obtain an initial spectral classification of such operators. The issue is easiest in one dimension.

Theorem 5.4.1 *Let H be the Friedrichs extension on $L^2(-1,1)$ of the symmetric degenerate elliptic operator defined initially on $C_c^\infty(-1,1)$ by*

$$Hf := -\frac{\mathrm{d}}{\mathrm{d}x}\left\{a(x)\frac{\mathrm{d}f}{\mathrm{d}x}\right\},$$

where the coefficient function $a(x)$ is C^1 on $(-1,1)$ and satisfies $a(x) \ge c(1-x^2)^\alpha$ for some $c > 0$, some $\alpha \in (0,1)$ and all $x \in (-1,1)$. Then $\mathrm{Spec}(H) \subseteq [\lambda, \infty)$, *where*

$$\lambda := \frac{c}{4}(1-\alpha)^2 > 0.$$

Proof Using the inequality $1 - x^2 \geq 1 - |x| \geq 0$ we have

$$\begin{aligned}
Q(f) &\geq c \int_{-1}^{1} (1 - |x|)^{\alpha} |f'(x)|^2 \mathrm{d}x \\
&\geq c \frac{(1-\alpha)^2}{4} \int_{-1}^{1} (1 - |x|)^{\alpha - 2} |f|^2 \mathrm{d}x \\
&\geq c \frac{(1-\alpha)^2}{4} \int_{-1}^{1} |f|^2 \mathrm{d}x
\end{aligned}$$

for all $f \in C_c^{\infty}(-1, 1)$. This bound remains valid for all f in the domain of the closure of Q, which equals $\mathrm{Dom}(H^{1/2})$. The theorem follows. □

Although the lower bound of the above theorem is not optimal, no such result is possible for $\alpha = 1$. The Legendre operator of Example 1.2.4 is the Friedrichs extension of its restriction to $C_c^{\infty}(-1, 1)$ and its smallest eigenvalue is 0.

We now consider the same problem in higher dimensions. In order to simplify the analysis, we shall only treat the case of operators acting on suitable spaces of functions on the unit ball in $\mathbf{R}^N$. However, entirely similar theorems can be proved for any bounded region in $\mathbf{R}^N$ with a smooth enough boundary. We need the following version of Hardy's inequality.

Lemma 5.4.2 *Let $\rho(s) \geq 0$ on $(0, b]$ and suppose that $\rho'(s) \leq 0$ for all $s \in (0, b]$. Let $-\infty < \alpha < 1$ and let $f \in C_c^{\infty}[0, b]$ vanish in a neighbourhood of 0. Then*

$$\frac{(1-\alpha)^2}{4} \int_0^b s^{\alpha-2} |f(s)|^2 \rho(s) \mathrm{d}s \leq \int_0^b s^{\alpha} |f'(s)|^2 \rho(s) \mathrm{d}s.$$

Proof The general case can be deduced from the case in which f is real-valued. If $\lambda := (1 - \alpha)/2$, then

$$\begin{aligned}
\int_0^b s^{\alpha} |f'|^2 \rho \, \mathrm{d}s &= \int_0^b s^{\alpha} \left| s^{\lambda} (s^{-\lambda} f)' + \lambda s^{-1} f \right|^2 \rho \, \mathrm{d}s \\
&\geq \lambda^2 \int_0^b s^{\alpha-2} |f|^2 \rho \, \mathrm{d}s + 2\lambda \int_0^b (s^{-\lambda} f)' s^{-\lambda} f \rho \, \mathrm{d}s \\
&= \lambda^2 \int_0^b s^{\alpha-2} |f|^2 \rho \, \mathrm{d}s + \lambda \left[(s^{-\lambda} f)^2 \rho \right]_0^b - \int_0^b (s^{-\lambda} f)^2 \rho' \, \mathrm{d}s \\
&\geq \lambda^2 \int_0^b s^{\alpha-2} |f|^2 \rho \, \mathrm{d}s.
\end{aligned}$$

□

Let b be a positive continuous function on the unit ball B in $\mathbf{R}^N$, and put $L^2 := L^2(B, b(x)\mathrm{d}^N x)$. Let $\{a_{i,j}(x)\}$ be a positive real symmetric matrix for each $x \in B$ and suppose that the coefficients are C^1 functions of x. Define $H : C_c^\infty(B) \to L^2$ by

$$Hf := -b^{-1} \sum_{i,j=1}^{N} \frac{\partial}{\partial x_i} \left\{ a_{i,j} \frac{\partial f}{\partial x_j} \right\},$$

so that

$$\langle Hf, f \rangle = \int_B \sum_{i,j=1}^{N} a_{i,j} \frac{\partial f}{\partial x_i} \frac{\partial \overline{f}}{\partial x_j} \mathrm{d}^N x \geq 0.$$

Finally let $\overline{H}$ denote the Friedrichs extension of H.

Theorem 5.4.3 *Suppose that the matrix inequality*

$$a(x) \geq c_1 (1 - |x|)^\alpha 1$$

holds for all $x \in B$, some $c_1 > 0$ and some $\alpha < 1$. Suppose also that

$$0 < b(x) \leq c_2 (1 - |x|)^{\alpha - 2}$$

for all $x \in B$ and some $c_2 > 0$. Then $\mathrm{Spec}(\overline{H}) \subseteq [\lambda, \infty)$, *where*

$$\lambda := \frac{c_1 (1 - \alpha)^2}{4 c_2} > 0.$$

Proof If $f \in C_c^\infty(B)$ then

$$\begin{aligned} \langle Hf, f \rangle &\geq c_1 \int_B (1 - |x|)^\alpha |\nabla f|^2 \mathrm{d}^N x \\ &\geq c_1 \int_S \int_0^1 (1 - r)^\alpha |f'(r)|^2 r^{N-1} \mathrm{d}r \mathrm{d}\omega, \end{aligned}$$

where $\mathrm{d}\omega$ is the surface measure suitably normalized on the unit sphere S of $\mathbf{R}^N$. An application of Lemma 5.4.2 with $s := 1 - r$ yields

$$\begin{aligned} \langle Hf, f \rangle &\geq \frac{c_1 (1 - \alpha)^2}{4} \int_S \int_0^1 (1 - r)^{\alpha - 2} |f|^2 r^{N-1} \mathrm{d}r \mathrm{d}\omega \\ &= \frac{c_1 (1 - \alpha)^2}{4} \int_B (1 - |x|)^{\alpha - 2} |f|^2 \mathrm{d}^N x \\ &\geq \frac{c_1 (1 - \alpha)^2}{4 c_2} \int_B |f|^2 b \, \mathrm{d}^N x \\ &= \lambda \|f\|_2^2. \end{aligned}$$

This inequality extends by continuity to all f in the domain of the form associated to the Friedrichs extension. □

If the coefficients $\{a_{i,j}(x)\}$ vanish more rapidly as $|x| \to 1$, then the conclusion of the above theorem is false.

Theorem 5.4.4 *Suppose that the matrix inequality*

$$0 \le a(x) \le c_1(1-|x|)1$$

holds for all $x \in B$. *Then* $0 \in \mathrm{Spec}(\overline{H})$.

Proof If f is a function of $s := 1-|x|$, then

$$\begin{aligned} Q(f) &= \int_B \sum_{i,j}^N a_{i,j} \frac{\partial f}{\partial x_i} \frac{\partial \overline{f}}{\partial x_j} \mathrm{d}^N x \\ &\le \int_B (1-|x|)|\nabla f|^2 \mathrm{d}^N x \\ &= \int_0^1 s|f'(s)|^2 \mathrm{d}s. \end{aligned}$$

Now let $0 < \delta < \varepsilon < 1$ and put

$$f_{\varepsilon,\delta}(s) := \begin{cases} \dfrac{\log(s/\delta)}{\log(\varepsilon/\delta)} & \text{if } \delta < s < \varepsilon \\ 0 & \text{if } 0 \le s \le \delta \\ 1 & \text{if } \varepsilon \le s \le 1. \end{cases}$$

A direct computation shows that

$$\int_0^1 s|f'_{\varepsilon,\delta}(s)|^2 \mathrm{d}s = \frac{1}{\log(\varepsilon/\delta)},$$

which converges to 0 if $\varepsilon \to 0$, $\delta \to 0$ and $\varepsilon/\delta \to \infty$. If we now put $\varepsilon := 1/2n$, $\delta := 1/4n^2$ and $f_n := f_{\varepsilon,\delta}$, then $0 \le f_n \le 1$,

$$\int_B |f_n|^2 b \,\mathrm{d}^N x \ge \int_{|x| \le \frac{1}{2}} b \,\mathrm{d}^N x > 0$$

and

$$\lim_{n\to\infty} Q(f_n) = 0.$$

If $g_n := f_n/\|f_n\|_2$ then we see that $\|g_n\|_2 = 1$ and $\lim_{n\to\infty} Q(g_n) = 0$, which implies the statement of the theorem. □

5.5 The biharmonic operator

When one considers small transverse vibrations of an elastic plate as opposed to an elastic membrane, the differential operator which describes the resonant frequencies is of fourth order rather than second order. A real-life problem of this type is to analyse the resonant frequencies (eigenvalues) of a rotating turbine blade; this can actually be done using a fourth order ordinary differential operator, with variable coefficients which incorporate the varying geometry of the blade from root to tip. A proper description of this problem is exceedingly complicated, and we content ourselves with the analysis of a simple approximate model.

Let Ω be a bounded region in $\mathbf{R}^N$ and let $H := \Delta^2$ with initial domain the subspace $C_c^\infty(\Omega)$ of $L^2(\Omega)$. This operator is clearly symmetric and the associated quadratic form is

$$Q(f) := \int_\Omega |\Delta f|^2 \mathrm{d}^N x,$$

which we take to have the same domain. We define $\overline{H}$ to be the Friedrichs extension of H; see Theorem 4.4.5.

We wish to emphasise that H is not the same operator as H_D^2, where H_D is minus the Dirichlet Laplacian of $L^2(\Omega)$. It can be shown that the classical boundary conditions appropriate to H are $f|_{\partial\Omega} = 0$ and $\partial f/\partial n|_{\partial\Omega} = 0$, where $\partial f/\partial n$ denotes the normal derivative of f, whereas those appropriate to H_D^2 are $f|_{\partial\Omega} = 0$ and $\Delta f|_{\partial\Omega} = 0$. We give a simple example to demonstrate that the two operators are distinct below.

In the following theorem we anticipate the general definition of $W_0^{1,2}(\Omega)$ given in Section 6.1. However, for the later results in the section the definition given in Example 4.4.4 suffices.

Theorem 5.5.1 *The operator $\overline{H}$ on $L^2(\Omega)$ has compact resolvent and quadratic form domain contained in* $\mathrm{Dom}(H_D^{1/2}) = W_0^{1,2}(\Omega)$.

Proof It follows by completing squares that

$$\int_\Omega |\Delta f|^2 \mathrm{d}^N x \geq \int_\Omega \{-2\overline{f}\Delta f - |f|^2\} \mathrm{d}^N x$$
$$= \int_\Omega \{2|\nabla f|^2 - |f|^2\} \mathrm{d}^N x$$

for all $f \in \mathbf{C}_c^\infty$. If Q and Q_D are the closed forms associated to $\overline{H}$ and $\overline{H_D}$ respectively, we deduce that

$$Q(f) \geq 2Q_D(f) - \|f\|^2$$

for all $f \in \mathrm{Dom}(\overline{H}^{1/2})$. This yields the second statement of the theorem, and the first is proved by using the variational formulae of Section 4.5. □

Example 5.5.2 We consider the simplest example possible in order to show that the operators $\overline{H}$ and $\overline{H_D}^2$ are distinct. Let $Hf := \mathrm{d}^4 f/\mathrm{d}x^4$ with domain the dense subspace C_c^∞ of $L^2(0,1)$. The eigenfunctions of $\overline{H}$ are solutions of the differential equation $\mathrm{d}^4 f/\mathrm{d}x^4 = \lambda f$ and must therefore be simple combinations of exponentials. This proves that every eigenfunction of $\overline{H}$ lies in $C^\infty[0,1]$, and we only have to show that the appropriate boundary conditions are different for the two operators. □

Lemma 5.5.3 *If* $f \in C^\infty[0,1]$, *then* $f \in \mathrm{Dom}(\overline{H}^{1/2})$ *if and only if* $f(0) = f'(0) = f(1) = f'(1) = 0$.

Proof We define $\tilde{Q}(f)$ for all $f \in C^\infty[0,1]$ by

$$\tilde{Q}(f) := \int_0^1 |f''|^2 \mathrm{d}x.$$

Suppose $f \in C^\infty[0,1]$ and that f and f' vanish at the ends of the interval. Let $\psi : \mathbf{R} \to [0,1]$ be smooth and satisfy $\psi(x) = 0$ if $x \le 1$ and $\psi(x) = 1$ if $x \ge 2$. If we define $\phi_n(x) := \psi(nx)\psi(n - nx)$, then $\phi_n \in \mathbf{C}_c^\infty(0,1)$, $0 \le \phi_n(x) \le 1$, $|\phi_n'(x)| \le c/x(1-x)$ and $|\phi_n''(x)| \le c/x^2(1-x)^2$ for some $c < \infty$, all $n \ge 1$ and all $0 < x < 1$. Now $f\phi_n \in \mathbf{C}_c^\infty$ and

$$\tilde{Q}(f - f\phi_n) \le \int_0^1 |f'' - f''\phi_n - 2f'\phi_n' - f\phi_n''|^2 \mathrm{d}x.$$

By combining the upper bounds on ϕ_n and its derivatives with the bounds $|f''(x)| \le c$, $|f'(x)| \le cx(1-x)$ and $|f(x)| \le cx^2(1-x)^2$, valid for all $x \in [0,1]$, we see that the integrand is bounded above uniformly with respect to x and n. Since the integrand converges pointwise to zero as $n \to \infty$, the dominated convergence theorem implies that $\lim_{n\to\infty} \tilde{Q}(f - f\phi_n) = 0$. Since $f\phi_n \in C_c^\infty(0,1)$ it follows that f lies in the domain of Q, which coincides with $\tilde{Q}$ on $C_c^\infty(0,1)$.

Conversely, suppose that $f \in \mathrm{Dom}(\overline{H}^{1/2})$. A combination of Lemma 5.3.3 and the definition of the Friedrichs extension shows that

$$\int_0^1 \frac{9|f(x)|^2}{16x^4} \mathrm{d}x \le Q(f) < \infty. \tag{5.5.1}$$

If in addition we know that $f \in C^\infty[0,1]$ then (5.5.1) implies that $f(0) =$

$f'(0) = 0$. The boundary conditions at $x = 1$ are proved by a similar method. □

Lemma 5.5.4 *The eigenvalues of* $\overline{H}$ *are of the form* $\lambda = \alpha^4$ *where* α *is any positive solution of*

$$\cos(\alpha)\cosh(\alpha) = 1. \tag{5.5.2}$$

Proof We first prove that every $\lambda > 0$ of the above form is an eigenvalue by writing down the corresponding eigenfunction. Given $\alpha > 0$, put

$$f(x) := \sin(\alpha x) - \sinh(\alpha x),$$
$$g(x) := \cos(\alpha x) - \cosh(\alpha x).$$

These functions both solve the differential equation and satisfy the boundary conditions at $x = 0$. The combination $f + \beta g$ satisfies the boundary conditions at $x = 1$ if and only if

$$\sin(\alpha) - \sinh(\alpha) + \beta\{\cos(\alpha) - \cosh(\alpha)\} = 0$$
$$\cos(\alpha) - \cosh(\alpha) - \beta\{\sin(\alpha) + \sinh(\alpha)\} = 0.$$

On eliminating β, we discover that this condition is equivalent to (5.5.2). The following graphs of $\phi(x) := \cos(x)$ and $\psi(x) := 1/\cosh(x)$ make it clear that the eigenvalues of $\overline{H}$ are numerically very close to $\{(2n+1)\pi/2\}^4$ for $n := 1, 2, \ldots$.

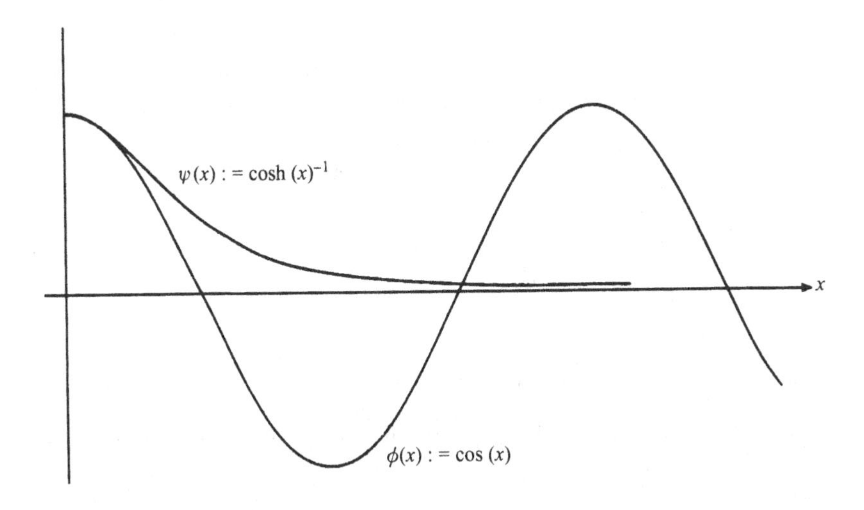

The proof that we have found all of the eigenvalues and eigenfunctions involves two issues. The first is that every weak solution of the eigenfunction equation is a combination of the four obvious exponentials. The second is that every such combination which satisfies the boundary conditions is of the form described above. Since this is only an example, we leave the reader to think about both of these questions. □

Exercises

5.1 Write down reformulations of Lemma 5.1.1 and 5.1.2 appropriate for complex Hilbert spaces.

5.2 Given $H = H^* \geq 0$ define e^{-Ht} as in Section 5.2. Prove that if $f \in \mathcal{H}$ then $f(t) := e^{-Ht}f$ is a norm continuous function for $t \geq 0$ and that it is norm differentiable for $t > 0$ with

$$\frac{df}{dt} = -H\{f(t)\}.$$

5.3 Let $g : [0, t] \to \mathcal{H}$ be norm continuous and suppose that g is norm differentiable on $(0, t)$ and satisfies

$$\frac{dg}{ds} = -H\{g(s)\}$$

for all $0 < s < t$. Prove that

$$h(s) := e^{-Hs}\{g(t-s)\}$$

has zero derivative for $s \in (0, t)$, and deduce that

$$g(t) = e^{-Ht}\{g(0)\}.$$

5.4 Continuing with the notation of Example 5.2.3, use the interpolation theorems of Section 3.6 to prove that the formula $T_t f = k_t * f$ defines a bounded linear operator from L^p to L^q for all $1 \leq p \leq q \leq \infty$, and obtain an upper bound on the norm of that operator.

5.5 Obtain an analogue of Theorem 5.2.4 for the operator $H := (-\Delta)^m$ acting on $L^2(\mathbf{R}^N)$.

5.6 Prove that

$$\int_0^1 \frac{|f|^2}{4x\log(x)^2}dx \leq \int_0^1 x|f'|^2 dx$$

for all $f \in C_c^\infty(0, 1)$. Hint: Replace x^λ in the proof of Lemma 5.3.1 by $|\log(x)|^\lambda$.

5.7 Use Theorem 5.3.5 to prove that if Ω is a convex region in $\mathbf{R}^2$, then

$$\frac{1}{4}\int_\Omega \frac{|f|^2}{d^2}\mathrm{d}^2x \le \int_\Omega |\nabla f|^2 \mathrm{d}^2x$$

for all $f \in C_c^\infty(\Omega)$. Deduce that $\mathrm{Spec}(H) \subseteq [\lambda, \infty)$, where $\lambda^{-1} := 4\,\mathrm{Inradius}(\Omega)^2$.

6
Dirichlet boundary conditions

6.1 Dirichlet boundary conditions

When one considers a differential operator acting in $L^2(\Omega)$, where Ω is a region in $\mathbf{R}^N$, it is normally necessary to specify boundary conditions. In this section we consider the easiest case, that of Dirichlet boundary conditions for a second order elliptic operator. Neumann and other boundary conditions will be considered in Chapter 7.

As well as being simple to treat, Dirichlet boundary conditions are directly relevant to a number of physical problems. These include heat flow in a medium whose boundary is kept at zero temperature, vibrations of an elastic membrane whose boundary is fixed, and the motion of a quantum particle which is confined to a region by the barrier associated with a large chemical potential.

We consider the operator on $L^2(\Omega)$ given formally by

$$Hf := -b(x)^{-1} \sum_{i,j=1}^{N} \frac{\partial}{\partial x_i} \left\{ a_{i,j}(x) \frac{\partial f}{\partial x_j} \right\}. \tag{6.1.1}$$

We assume initially that $a(x) := \{a_{i,j}(x)\}$ is a real symmetric matrix-valued function of x which is continuously differentiable up to and including the boundary $\partial\Omega$; more precisely $a_{i,j}(x)$ and their first partial derivatives are continuous in x and have continuous extensions to the closure $\overline{\Omega}$. We also assume that the matrices $a(x)$ are uniformly positive and bounded in the sense that there exists a constant $c \geq 1$ such that

$$c^{-1} 1 \leq a(x) \leq c\, 1 \tag{6.1.2}$$

in the sense of matrices, for all $x \in \Omega$. We finally assume that the real-valued function $b(x)$ is continuous up to and including the boundary of Ω, and that

$$c^{-1} \leq b(x) \leq c \tag{6.1.3}$$

for all $x \in \Omega$. The conditions on $a(x)$ and $b(x)$ will be relaxed later.

There are several different reasons for studying operators with variable coefficients. The first is that one may wish to study vibrations or heat flow in a material of variable thickness or variable composition. The second is that the study of such problems for a body in the form of a curved surface involves the use of curvilinear coordinates even if the body has constant thickness and composition. More generally, the analysis of Laplace–Beltrami operators on Riemannian manifolds clearly depends upon the use of local coordinates and variable coefficients.

The reader will observe that the operator given by (6.1.1) is not that associated with the symbol

$$a(x,y) := b(x)^{-1} \sum_{i,j=1}^{N} a_{i,j}(x) y_i y_j$$

according to the procedure of Section 3.5. Operators written in the form of (1.1.1) are said to be in the divergence form, which is particularly suitable for spectral analysis. When one considers operators with non-smooth coefficients, the difference between the divergence and non-divergence forms becomes critical, and one actually has two different if related theories.

The operator H acts in the Hilbert space $L^2(\Omega, b(x)\mathrm{d}^N x)$. As a linear space this coincides with $L^2(\Omega, \mathrm{d}^N x)$, but it has a different norm and inner product. We define the initial domain $C_0^\infty(\overline{\Omega})$ of the operator H as follows. We say that $f \in C^\infty(\overline{\Omega})$ if f is a smooth function on Ω all of whose partial derivatives can be extended continuously to $\overline{\Omega}$. For f to be in $C_0^\infty(\overline{\Omega})$ we impose the extra requirement that $f(x) = 0$ for all $x \in \partial\Omega$. This condition incorporates the notion of Dirichlet boundary conditions.

Lemma 6.1.1 *Under the above conditions the operator H is symmetric and non-negative. The associated quadratic form*

$$Q(f,g) := \int_\Omega \sum_{i,j=1}^{N} a_{i,j}(x) \frac{\partial f}{\partial x_i} \frac{\partial \overline{g}}{\partial x_j} \mathrm{d}^N x \tag{6.1.4}$$

is closable on the domain $C_0^\infty(\overline{\Omega})$ and the domain of the closure is independent of the particular coefficients b, $\{a_{i,j}\}$ chosen.

Proof If $f, g \in C_0^\infty(\overline{\Omega})$ then integration by parts (Gauss' theorem) establishes that

$$\langle Hf, g\rangle = Q(f,g) = \langle f, Hg\rangle.$$

Thus H is symmetric. Since the matrices $\{a_{i,j}(x)\}$ are non-negative, the form Q is non-negative. Theorem 4.4.5 implies that the form Q is closable. The bounds (6.1.2) and (6.1.3) imply that

$$c^{-1}|||f|||^2 \le |||f|||_1^2 := \int_\Omega \Big\{ b(x)|f|^2 + \sum_{i,j=1}^N a_{i,j}(x) \frac{\partial f}{\partial x_i}\frac{\partial \overline{f}}{\partial x_j}\Big\} \mathrm{d}^N x \le c|||f|||^2, \tag{6.1.5}$$

where

$$|||f|||^2 := \int_\Omega \{|f|^2 + |\nabla f|^2\}\mathrm{d}^N x \tag{6.1.6}$$

is independent of the coefficients b, $a_{i,j}$. The completions of $C_0^\infty(\overline{\Omega})$ for the two norms are equal because the norms are comparable. Compare Corollary 4.4.3, where, however, there are two forms but only one Hilbert space. □

From now on we shall use the symbol H, or H_D if we particularly want to emphasise the choice of Dirichlet boundary conditions, to refer to the self-adjoint operator associated with the closure of the quadratic form above by the use of Theorem 4.4.5. Thus H_D is the Friedrichs extension of the operator defined initially on $C_0^\infty(\overline{\Omega})$. Under stronger smoothness conditions on the coefficients of H and on $\partial\Omega$ than those stated above, it is actually the case that H has a complete orthonormal set of eigenfunctions, all of which lie in the domain $C_0^\infty(\overline{\Omega})$. However, we shall not prove this fact in the present book. There are three reason for this. The first is that the statement is not true under the weaker conditions of this chapter. The second is that we are mainly interested in estimating the eigenvalues, and for this purpose it is more useful to study the quadratic forms than the operators themselves. The final reason (and perhaps the one of most importance for this text) is that the proof of the above statement is somewhat tedious!

Before proceeding any further we have to define the notion of weak (or distributional) derivatives for functions defined on a region Ω in $\mathbf{R}^N$. This is rather similar to what we did in Section 3.4, except that we can no longer use Schwartz space or Fourier transforms. We start by defining $C_c^\infty(\Omega)$ to be the space of smooth functions with compact supports contained in Ω. Ignoring questions of continuity as before, we define a distribution to be a linear functional $\phi : C_c^\infty(\Omega) \to \mathbf{C}$. If g is a function on Ω which is integrable when restricted to every compact subset of Ω, then g determines a distribution ϕ by means of the formula

$$\phi(f) := \int_\Omega f(x)g(x)\mathrm{d}^N x. \tag{6.1.7}$$

If α is any multi-index, the weak derivative $D^\alpha\phi$ of the distribution ϕ is defined as in Section 3.4 by

$$(D^\alpha\phi)(f) := (-1)^{|\alpha|}\phi(D^\alpha f).$$

One can check that this is compatible with the usual notion of derivative if ϕ is associated with a smooth function on Ω by means of the formula (6.1.7). If g is a smooth function on Ω, then we define the product $g\phi$ to be the distribution $(g\phi)(f) := \phi(gf)$.

Given any region Ω in $\mathbf{R}^N$ we define the Sobolev space $W^{1,2}(\Omega)$ to be the space of all functions $f \in L^2(\Omega)$ whose weak derivatives $\partial_i f := \partial f/\partial x_i$ all lie in $L^2(\Omega)$. Thus $W^{1,2}(\Omega)$ is the domain of the gradient operator ∇ from $L^2(\Omega)$ into $L^2(\Omega) \oplus ... \oplus L^2(\Omega)$ defined by

$$\nabla f(x) := \partial_1 f \oplus ... \oplus \partial_N f.$$

We define an inner product on $W^{1,2}(\Omega)$ by

$$\langle f, g\rangle_1 := \int_\Omega \{f(x)\overline{g(x)} + \nabla f(x)\cdot\overline{\nabla g(x)}\}\mathrm{d}^N x, \tag{6.1.8}$$

the associated norm being given by (6.1.6), but for the enlarged class of f.

It is true, but not obvious, that if $\Omega = \mathbf{R}^N$ then our definition of $W^{1,2}(\Omega)$ coincides with that of Section 3.7. The reason is that we are using two slightly different definitions of weak derivative. This loose thread is tied in Exercise 6.3.

Lemma 6.1.2 *For any choice of the region $\Omega \subseteq \mathbf{R}^N$ the space $W^{1,2}(\Omega)$ is a Hilbert space with respect to the inner product (6.1.8).*

Proof A little thought shows that this is equivalent to the statement that ∇ is a closed linear operator. Let $\{f_n\}_{n=1}^\infty$ be a sequence in $W^{1,2}(\Omega)$

and let $f, g_1, ..., g_N \in L^2(\Omega)$. Assume that $\lim_{n\to\infty} \|f_n - f\|_2 = 0$ and that $\lim_{n\to\infty} \|\partial_i f_n - g_i\|_2 = 0$ for all $1 \leq i \leq N$. If $\phi \in C_c^\infty(\Omega)$, then

$$\begin{aligned}\langle g_i, \phi\rangle &= \lim_{n\to\infty} \langle \partial_i f_n, \phi\rangle \\ &= -\lim_{n\to\infty} \langle f_n, \partial_i \phi\rangle \\ &= -\langle f, \partial_i \phi\rangle \\ &= \langle \partial_i f, \phi\rangle.\end{aligned}$$

Therefore $g_i = \partial_i f$ and $f \in \mathrm{Dom}\nabla$ with $\nabla f = (g_1, ..., g_N)$. □

We now define the subspace $W_0^{1,2}(\Omega)$ of $W^{1,2}(\Omega)$ to be the closure of the subspace $C_c^\infty(\Omega)$ for the norm $\|\| \ \|\|$. Typically $W_0^{1,2}(\Omega)$ and $W^{1,2}(\Omega)$ are not equal. Our next lemma gives a first indication of the importance of the Sobolev space $W_0^{1,2}(\Omega)$. It also gives the first of several sufficient conditions for a function to lie in $W_0^{1,2}(\Omega)$.

Lemma 6.1.3 *The domain of the closure of the quadratic form Q defined in (6.1.4) is precisely $W_0^{1,2}(\Omega)$. Thus $W_0^{1,2}(\Omega)$ equals the domain of $H_D^{1/2}$ and contains $C_0^\infty(\overline{\Omega})$.*

Proof Since $C_c^\infty(\Omega) \subseteq C_0^\infty(\overline{\Omega})$ we see that

$$W_0^{1,2}(\Omega) \subseteq \mathrm{Dom}(\overline{Q}) \subseteq W^{1,2}(\Omega).$$

We shall show that $C_c^\infty(\Omega)$ is dense in $C_0^\infty(\overline{\Omega})$ for the norm $\|\| \ \|\|$. This will establish that $\mathrm{Dom}(\overline{Q}) \subseteq W_0^{1,2}(\Omega)$, and complete the proof of the lemma. By taking real and imaginary parts, it is sufficient to treat the case in which the functions concerned are real-valued.

Given $\varepsilon > 0$ let $F_\varepsilon : \mathbf{R} \to \mathbf{R}$ be a smooth function satisfying

(1) $F_\varepsilon(x) = x$ if $|x| \geq 2\varepsilon$
(2) $F_\varepsilon(x) = 0$ if $|x| \leq \varepsilon$
(3) $|F_\varepsilon(x)| \leq |x|$ for all $x \in \mathbf{R}$
(4) $0 \leq F_\varepsilon'(x) \leq 3$ for all $x \in \mathbf{R}$.

If $f \in C_0^\infty(\overline{\Omega})$, put $f_\varepsilon(x) := F_\varepsilon(f(x))$ for all $x \in \Omega$. It follows immediately from the definition that f_ε is a smooth function on Ω which vanishes in a neighbourhood of $\partial\Omega$. Moreover, $|f_\varepsilon(x)| \leq |f(x)|$ and $\lim_{\varepsilon\to 0} f_\varepsilon(x) = f(x)$ for all $x \in \Omega$. The dominated convergence theorem now implies that $\|f - f_\varepsilon\|_2 \to 0$ as $\varepsilon \to 0$. Also $|\nabla f_\varepsilon(x)| \leq 3|\nabla f(x)|$ and $\lim_{\varepsilon\to 0} \nabla f_\varepsilon(x) = \nabla f(x)$ for all $x \in \Omega$ such that $f(x) \neq 0$.

Although it is often the case that $A := \{x : f(x) = 0\}$ is a Lebesgue null set, this need not be true. However, $B := \{x : f(x) = 0 \text{ and } \nabla f(x) \neq 0\}$

is a hypersurface of codimension 1 by the implicit function theorem, and so must be a null set. The dominated convergence theorem establishes that

$$\lim_{\varepsilon\to 0} \|\nabla f - \nabla f_\varepsilon\|_2^2 = \int_A |\nabla f|^2 \mathrm{d}^N x = \int_B |\nabla f|^2 \mathrm{d}^N x = 0$$

and this completes the proof. □

We are now able to give a far-reaching generalization of the above construction. Although we are mostly interested in differential operators with piecewise smooth coefficients and regions with piecewise smooth boundaries, there is no extra cost in formulating our ideas below in much greater generality. This is of particular value in the study of the bulk properties of materials with randomly distributed impurities. We suppose that Ω is any region in $\mathbf{R}^N$ and that $\{a_{i,j}(x)\}$ is a real symmetric matrix depending measurably upon the variable $x \in \Omega$ and satisfying (6.1.2). We also suppose that $b(x)$ is a positive measurable function on Ω satisfying (6.1.3). The definition (6.1.1) of the operator H no longer makes sense in any obvious way because one does not necessarily even have $C_c^\infty(\Omega) \subseteq \mathrm{Dom}(H)$. Nevertheless we can associate a self-adjoint operator to the expression in the following way.

Theorem 6.1.4 *Under the conditions just stated, the quadratic form Q defined by (6.1.4) is closed on the domain $W_0^{1,2}(\Omega)$ in the Hilbert space $L^2(\Omega, b(x)\mathrm{d}^N x)$. There exists a non-negative self-adjoint operator H_D on $L^2(\Omega, b(x)\mathrm{d}^N x)$ associated to the form, in such a way that*

$$\langle H_D^{1/2} f, H_D^{1/2} g\rangle = Q(f, g)$$

for all $f, g \in \mathrm{Dom}(H_D^{1/2}) = W_0^{1,2}(\Omega)$.

Proof It follows from (6.1.5) that the form is indeed closed on $W_0^{1,2}(\Omega)$, and the other statements of the theorem now follow from Theorem 4.4.2. □

The fact that the domain of $H_D^{1/2}$ is independent of the coefficients $a_{i,j}(x)$ and $b(x)$ subject to (6.1.2) and (6.1.3) is extremely important for the study of operators with measurable coefficients.

There is an alternative approach to the analysis of operators with discontinuous coefficients. This is to carry out a regularisation procedure which replaces the operator H and region Ω by operators H_ε with smooth coefficients and regions Ω_ε with smooth boundaries. Although the qualitative aspects of the analysis of H_ε and Ω_ε are considerably simpler for each $\varepsilon > 0$, the systematic quantitative investigation of the limit $\varepsilon \to 0$ involves the same kind of issue as our more direct approach.

Although the Sobolev space $W_0^{1,2}(\Omega)$ is very important in our analysis below, it is not very easy to see whether a given function lies in the space. The following criterion is often useful.

Theorem 6.1.5 *Let $f : \mathbf{R}^N \to \mathbf{R}$ be a continuous function which vanishes outside a bounded region Ω and satisfies the Lipschitz condition*

$$|f(x) - f(y)| \le c|x - y| \tag{6.1.9}$$

for some $c < \infty$ and all $x, y \in \mathbf{R}^N$. Then $f \in W_0^{1,2}(\Omega)$.

Proof We break the proof up into four parts. The main idea is that functions of the above type have bounded weak first derivatives.

Part (1) We first show that if f is a function satisfying (6.1.9) whose support is a compact subset K of Ω then $f \in W^{1,2}(\mathbf{R}^N)$ and

$$\begin{aligned}|||f|||^2 :&= \int_{\mathbf{R}^N} \{|\nabla f|^2 + |f|^2\} \mathrm{d}^N x \\ &\le |K| c^2 \{N + \operatorname{diam}(K)^2\} < \infty,\end{aligned}$$

where $|K|$ denotes the Lebesgue measure of K and $\operatorname{diam}(K)$ denotes its diameter.

Let k_s be a standard mollifier, as defined in Section 3.2, so that $k_s(x) = s^{-N} k(s^{-1}x)$ for all $x \in \mathbf{R}^N$ and all $s > 0$. If $p_s := k_s * f$, then

$$K_s := \operatorname{supp}(p_s) \subseteq \{x : \operatorname{dist}(x, K) \le s\}$$

and $|K_s| \to |K|$ as $s \to 0$. By differentiating

$$p_s(x) = \int_{\mathbf{R}^N} k_s(x - y) f(y) \mathrm{d}^N y$$

under the integral sign we see that $p_s \in C_c^\infty(\mathbf{R}^N)$. On the other hand from

$$p_s(x) - p_s(y) = \int_{\mathbf{R}^N} \{f(x - z) - f(y - z)\} k_s(z) \mathrm{d}^N z$$

we can obtain the bound

$$|p_s(x) - p_s(y)| \le c|x - y|$$

for all $x, y \in \mathbf{R}^N$, with the same constant c as before. Since p_s is smooth this bound implies that $|\nabla p_s(x)| \le c$ for all $x \in \mathbf{R}^N$. Elementary estimates now imply that

$$|||p_s|||^2 \le |K_s| c^2 \{1 + \mathrm{diam}(K_s)^2\}$$

for all $s > 0$.

Now for any $p \in W^{1,2}(\mathbf{R}^N)$ we have

$$\begin{aligned} |||p|||^2 &= \|p\|_2^2 + \sum_{r=1}^{N} \left\| \frac{\partial p}{\partial x_r} \right\|_2^2 \\ &= \int_{\mathbf{R}^N} (1 + |y|^2) |\hat{p}(y)|^2 \mathrm{d}^N y. \end{aligned}$$

We apply this to p_s, using the identity $\hat{p}_s(y) = \hat{f}(y) h(sy)$, where

$$h(y) := \int_{\mathbf{R}^N} k(x) \mathrm{e}^{-ix \cdot y} \mathrm{d}^N x$$

is continuous with $h(0) = 1$ and $|h(x)| \le 1$ for all $x \in \mathbf{R}^N$. Fatou's lemma implies that

$$\begin{aligned} \int_{\mathbf{R}^N} |\hat{f}(y)|^2 (1 + |y|^2) \mathrm{d}^N y &\le \liminf_{s \to 0} \int_{\mathbf{R}^N} |\hat{f}(y)|^2 h(sy)^2 (1 + |y|^2) \mathrm{d}^N y \\ &= \liminf_{s \to 0} |||p_s|||^2 \\ &\le |K| c^2 \{N + \mathrm{diam}(K)^2\} < \infty. \end{aligned}$$

Hence $f \in W^{1,2}(\mathbf{R}^N)$ and

$$|||f|||^2 \le |K| c^2 \{N + \mathrm{diam}(K)^2\}.$$

Part (2) The dominated convergence theorem implies that

$$\lim_{s \to 0} |||f - p_s|||^2 = \lim_{s \to 0} \int_{\mathbf{R}^N} |\hat{f}(y)|^2 |1 - h(sy)|^2 (1 + |y|^2) \mathrm{d}^N y = 0.$$

The support of each p_s is slightly larger than the support of f, so p_s does not in general lie in $C_c^\infty(\Omega)$, and we need a further idea to complete the proof.

Part (3) It is possible, although unlikely, that $K := \mathrm{supp}(f)$ has much larger Lebesgue measure than $V := \{x : f(x) \neq 0\}$. In this case it is important that we can improve the above estimate of $|||f|||$. To do this, let $f_\varepsilon(x) := \phi_\varepsilon(f(x))$ for all $x \in V$ where

$$\phi_\varepsilon(s) = \begin{cases} 0 & \text{if } |s| \le \varepsilon, \\ s & \text{if } |s| \ge 2\varepsilon, \\ 2(s - \varepsilon) & \text{if } \varepsilon < s < 2\varepsilon, \\ 2(s + \varepsilon) & \text{if } -2\varepsilon < s < -\varepsilon. \end{cases}$$

The function f_ε has support in $S_\varepsilon := \{x : \mathrm{dist}(x, \mathbf{R}^N \backslash V) \geq \varepsilon/c\}$, and

$$|f_\varepsilon(x) - f_\varepsilon(y)| \leq 2|f(x) - f(y)| \leq 2c|x - y|$$

for all $x, y \in \mathbf{R}^N$. Part (1) shows that $f_\varepsilon \in W^{1,2}(\mathbf{R}^N)$ but also establishes the bound

$$|||f_\varepsilon|||^2 \leq 4|S_\varepsilon|c^2\{N + \mathrm{diam}(V)^2\}.$$

A further application of Fatou's lemma to the Fourier transforms of f_ε and f now shows that

$$|||f|||^2 \leq 4|V|c^2\{N + \mathrm{diam}(V)^2\}.$$

Part (4) Now let f be the function considered in the statement of the theorem. By applying Parts (1) and (2) to f_ε we see that $f_\varepsilon \in W_0^{1,2}(\Omega)$. Now $g_\varepsilon := f - f_\varepsilon = \psi_\varepsilon(f(x))$ where

$$\psi_\varepsilon(s) := \begin{cases} 0 & \text{if } |s| \geq 2\varepsilon, \\ s & \text{if } |s| \leq \varepsilon, \\ 2\varepsilon - s & \text{if } \varepsilon < s < 2\varepsilon, \\ -s - 2\varepsilon & \text{if } -2\varepsilon < s < -\varepsilon. \end{cases}$$

The set L_ε on which g_ε is non-zero is contained in $\{x : 0 < |f(x)| < 2\varepsilon\}$ and this set decreases to the empty set as $\varepsilon \to 0$. Part (3) of this proof establishes that $g_\varepsilon \in W^{1,2}(\mathbf{R}^N)$ with

$$|||g_\varepsilon|||^2 \leq 4|L_\varepsilon|c^2\{N + \mathrm{diam}(L_\varepsilon)^2\}.$$

Therefore

$$\lim_{\varepsilon \to 0} |||f - f_\varepsilon||| = \lim_{\varepsilon \to 0} |||g_\varepsilon||| = 0$$

and $f \in W_0^{1,2}(\Omega)$. □

Theorem 6.1.6 *Let $f \in W^{1,2}(\Omega)$ and suppose that $f(x) = 0$ almost everywhere outside a compact subset K of Ω. Then $f \in W_0^{1,2}(\Omega)$.*

Proof Let $f_s := k_s * f$ where k_s is the standard mollifier. The same type of argument as in Theorem 6.1.5 establishes that $f_s \in C_c^\infty(\Omega)$ for small enough $s > 0$. We also have

$$\lim_{s \to 0} |||f_s - f||| = 0$$

by a modification of the argument of Theorem 6.1.5. □

6.2 The Dirichlet Laplacian

In this section we consider the simplest case of the above theory, when $a_{i,j}(x) = \delta_{i,j}$ and $b(x) = 1$ for all $x \in \Omega$, so that the operator H_0 is minus the Laplacian acting on the space $L^2(\Omega, \mathrm{d}^N x)$. We assume Dirichlet boundary conditions in the sense that the quadratic form in question is taken to have domain equal to $W_0^{1,2}(\Omega)$. Our analysis depends heavily upon the solution of the following particular case.

Lemma 6.2.1 *Let Ω be the cube*

$$\{x = (x_1, ..., x_N) : 0 < x_i < a \text{ for all } 1 \le i \le N\}.$$

Then the functions

$$f_n(x) := \left(\frac{2}{a}\right)^{N/2} \prod_{i=1}^{N} \sin(\pi n_i x_i / a)$$

parametrised by the multi-index of positive integers $n := \{n_1, ..., n_N\}$, form a complete orthonormal set of eigenfunctions of the operator H_0 with corresponding eigenvalues

$$\lambda_n := \pi^2 a^{-2}(n_1^2 + ... + n_N^2).$$

The operator H_0 has compact resolvent.

Proof Each of the above functions lies in $C_0^\infty(\overline{\Omega})$, and Lemma 6.1.3 then implies that they lie in $W_0^{1,2}(\Omega)$. Now it is elementary that $-\Delta f_n = \lambda_n f_n$ in the classical sense of differential equations. The proof that that f_n lie in $\mathrm{Dom}(H_0)$ and are eigenfunctions of H_0 uses Lemma 4.4.1. Since the f_n form a complete orthonormal set in $L^2(\Omega)$ by Fourier analysis, there are no other eigenvalues apart from those listed. If we relabel the eigenvalues in increasing order, repeating them according to multiplicity, then they diverge to ∞. The last statement of the theorem is now a consequence of Theorem 4.2.3. □

The eigenvalues of the above operator may have multiple degeneracies. The following diagram shows some of the level curves of the eigenfunction

$$f(x, y) := \sin(5\pi x)\sin(5\pi y) + 0.6 \sin(7\pi x)\sin(\pi y)$$

associated with $(0, 1)^2$.

Corollary 6.2.2 *Let λ_n denote the eigenvalues of the above cube written in increasing order and repeated according to multiplicity. Then there exists a constant $c > 0$ such that*

$$\lim_{n\to\infty} \left(\lambda_n n^{-2/N}\right) = c.$$

Proof Let

$$M(s) := \#\{(n_1, ..., n_N) : \lambda_{n_1,...,n_N} \le s\}$$

for all $s \ge 0$. By comparing $\lambda_{n_1,...,n_N}$ with the distance from 0 of a point in $\mathbf{Z}^N$, and using the formula for the volume of a ball in $\mathbf{R}^N$ of radius $s^{1/2}$, we see that

$$k := \lim_{s\to\infty} \left\{M(s)s^{-N/2}\right\}$$

exists and is positive. The corollary follows upon observing that $M(\lambda_n) = n$. □

One of the central ideas for treating more general regions than the above cube is to use a monotonicity property of eigenvalues.

Theorem 6.2.3 *Let Ω be a bounded region in $\mathbf{R}^N$, and let H_0 be minus the Dirichlet Laplacian, acting on $L^2(\Omega)$. Then H_0 has empty essential spectrum and compact resolvent. Let $\lambda_n(\Omega)$ denote the nth eigenvalue of*

H_0. Then $\lambda_n(\Omega)$ is a monotonically decreasing function of the region, and is continuous from inside in the sense that for any increasing sequence of regions Ω_m with union equal to Ω we have

$$\lim_{m\to\infty} \lambda_n(\Omega_m) = \lambda_n(\Omega)$$

for all $n \geq 1$.

Proof We define

$$Q_0(f) := \int_\Omega |\nabla f(x)|^2 \mathrm{d}^N x \tag{6.2.1}$$

for all $f \in C_c^\infty(\Omega)$. We define

$$\lambda_n(\Omega) := \inf\{\lambda(L) : L \subseteq C_c^\infty(\Omega) \text{ and } \dim(L) = n\},$$

where

$$\lambda(L) := \sup\{Q_0(f) : f \in L \text{ and } \|f\| \leq 1\}$$

for all finite-dimensional linear subspaces of $C_c^\infty(\Omega)$. If $\Omega \subseteq \Omega'$, then $\lambda_n(\Omega) \geq \lambda_n(\Omega')$ since it is obtained by taking the infimum over a smaller class of linear subspaces. If Ω' is a cube then an application of Corollary 6.2.2 shows that $\lim_{n\to\infty}\{\lambda_n(\Omega)\} = \infty$. The variational theory of Section 4.5 now implies that H_0 has compact resolvent, and that $\lambda_n(\Omega)$ are the eigenvalues of H_0, written in increasing order and repeated according to multiplicity.

Now suppose that we wish to approximate the eigenvalue $\lambda_n(\Omega)$ to within an error $\varepsilon > 0$ by choosing a region $\Omega' \subseteq \Omega$. There exists an n-dimensional subspace L of $C_c^\infty(\Omega)$ such that $\lambda(L) \leq \lambda_n(\Omega) + \varepsilon$. There exists a compact subset K of Ω such that $\mathrm{supp}(f) \subseteq K$ for all $f \in L$. If Ω' is any open set such that $K \subseteq \Omega' \subseteq \Omega$, then it follows that

$$\lambda_n(\Omega) \leq \lambda_n(\Omega') \leq \lambda_n(\Omega) + \varepsilon.$$

This immediately implies the stated limit property. □

Although this theorem seems very satisfactory, it opens up several further questions which are far from answered at the present date. Under what conditions can one assert that the eigenvalues converge to the expected limit for a decreasing sequence of regions? Can one obtain precise information about the asymptotic rate of convergence in fairly general circumstances for increasing or decreasing sequences of regions? What are the answers to the corresponding questions for Neumann and other boundary conditions?

There are extremely few regions for which one can write down explicit expressions for the eigenvalues, and a large body of numerical analysis is devoted to finding approximate solutions to questions of this type. The task of analysts is to find new abstract techniques for the solution of problems, and also to obtain quantitative and qualitative information which helps to support numerical procedures. Among the exactly soluble problems we must mention the following.

Example 6.2.4 Let B_R denote the ball in $\mathbf{R}^2$ with centre 0 and radius $R > 0$. We may use the rotational group to decompose $L^2(B_R)$ into orthogonal linear subspaces $\{L_n\}_{n\in\mathbf{Z}}$, consisting of functions of the form

$$f(r\cos\theta, r\sin\theta) = g(r)\mathrm{e}^{in\theta},$$

where

$$\|g\|^2 := \|f\|_2^2 = 2\pi\int_0^R |g(r)|^2 r\mathrm{d}r.$$

This also yields a decomposition of $W_0^{1,2}(B_R)$ into linear subspaces of the same form, where

$$Q_n(g) := Q(f) = 2\pi\int_0^R \left\{|g'(r)|^2 + \frac{n^2}{r^2}|g(r)|^2\right\} r\mathrm{d}r$$

for all $f \in L_n$. Since the operator H_0 commutes with rotations, it maps each of these subspaces into itself, and its spectral behaviour can be analysed in each subspace independently.

Within the subspace L_n the operator H_0 reduces to the ordinary differential operator

$$K_n g(r) := -\frac{1}{r}\frac{\mathrm{d}}{\mathrm{d}r}\left(r\frac{\mathrm{d}g}{\mathrm{d}r}\right) + \frac{n^2}{r^2}g(r)$$

associated with the quadratic form Q_n. This operator acts on the Hilbert space $\mathscr{H}$ of functions $g : [0, R] \to \mathbf{C}$ such that $\|g\| < \infty$ and is associated with Dirichlet boundary conditions at $r = R$. The eigenfunctions of this operator are of the form

$$J_n(j_{n,s}r/R)\mathrm{e}^{in\theta},$$

where $s = 1, 2, ...$ and $j_{n,s}$ is the sth zero of the Bessel function J_n. The corresponding eigenvalue is $j_{n,s}^2/R^2$. The first ten of these eigenvalues are listed in increasing order for the unit disc in the Table 6.1.

Table 6.1. *Eigenvalues of the Unit Disc*

Name	Formula	Value
λ_1	$j_{0,1}^2$	5.784
λ_2	$j_{1,1}^2$	14.684
λ_3	$j_{-1,1}^2$	14.684
λ_4	$j_{2,1}^2$	26.378
λ_5	$j_{-2,1}^2$	26.378
λ_6	$j_{0,2}^2$	30.470
λ_7	$j_{3,1}^2$	40.704
λ_8	$j_{-3,1}^2$	40.704
λ_9	$j_{1,2}^2$	49.224
λ_{10}	$j_{-1,2}^2$	49.224

One point which can be seen clearly from the table is that there is little connection between the natural parametrisation of the eigenvalues in terms of the integers n and s, and the parametrisation given by writing them in monotonically increasing order. The latter choice is relevant to variational estimation procedures, but does not respect analytic considerations arising from group symmetry.

Without reference to the numerical values one can observe that the quadratic form Q_0 is smaller than all of the other quadratic forms, so the smallest eigenvalue of this form is also the smallest eigenvalue of the whole operator H_0. The smallest eigenvalue of the operator H_0 is of multiplicity 1, and is associated with a rotationally invariant eigenfunction which is positive for $|x| < R$ but zero for $|x| = R$. □

Our next example describes a phenomenon which is generic for the Laplacian, namely that the rate of decay of an eigenfunction at any particular point of the boundary depends upon the local geometry at that point. At vertices which point outwards eigenfunctions decay faster than linearly, while at vertices which point inwards they decay more slowly than linearly.

Example 6.2.5 Let $0 < \beta < 2\pi$ and let Ω be the sector

$$\{(r\cos\theta, r\sin\theta) : 0 < r < 1 \text{ and } 0 < \theta < \beta\}.$$

Although this region is not rotationally invariant, we can still analyse the Dirichlet Laplacian on it by decomposing $L^2(\Omega, \mathrm{d}^2x)$ into subspaces L_n of functions of the form

$$f(r\cos\theta, r\sin\theta) = g(r)\sin(\pi n\theta/\beta),$$

where $1 \le n < \infty$. The smallest eigenvalue in this case corresponds to an eigenfunction in the subspace L_1. The eigenfunction is given explicitly in terms of Bessel functions by

$$f(x, y) = J_{\pi/\beta}(c_{R,\beta}r)\sin(\pi\theta/\beta).$$

Once again the eigenfunction is positive in Ω and vanishes on the boundary $\partial\Omega$. However, the gradient of the eigenfunction is only bounded near the origin if $\beta \le \pi$.

The following diagram shows level curves of the above eigenfunction in the case $\beta = 2\pi$. In this particular case the eigenfunction has the simple expression

$$f(x, y) = c_1\frac{\sin(\pi r)}{r}(r - x)^{1/2}.$$

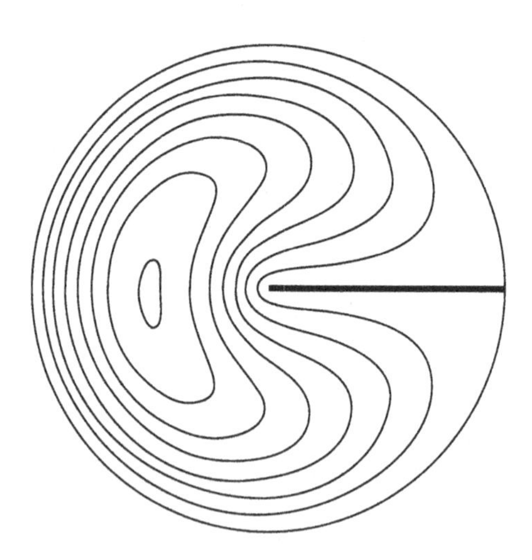

It is worth commenting that the eigenfunction f lies in $L^p(\Omega)$ for all $1 \le p \le \infty$ and that Δf lies in $L^p(\Omega)$ for all $1 \le p \le \infty$. However, assuming $\pi < \beta \le 2\pi$, ∇f only lies in $L^p(\Omega)$ if

$$1 \le p < \frac{2\beta}{\beta - \pi}.$$

The same result holds for all other eigenfunctions in the subspace L_1; eigenfunctions in other L_n spaces are better behaved in this respect. A reader who has had a little contact with elliptic regularity theorems may be puzzled by these facts. The point is that elliptic regularity theorems are either local statements referring to small neighbourhoods of points inside Ω, or global statements depending upon smoothness of the boundary. When considering regions with piecewise smooth boundaries it is quite possible for Δf to be more regular near the boundary than ∇f. □

Although we are avoiding the use of elliptic regularity theorems, we do at least indicate what one can do with them for the Laplacian. If Ω is a region in $\mathbf{R}^N$ we say that a function $f : \Omega \to \mathbf{C}$ lies in $W^{n,2}_{\text{loc}}(\Omega)$ if $f\phi \in H^n := W^{n,2}(\mathbf{R}^N)$ for all $\phi \in C_c^\infty(\Omega)$. We say that a vector-valued function lies in one of these spaces if each of its components does so.

Theorem 6.2.6 *If $f \in L^2_{\text{loc}}(\Omega)$ and $g \in W^{n,2}_{\text{loc}}(\Omega)$ are related by the identity*

$$-\Delta f = \lambda f + g$$

interpreted in the weak sense, for some $\lambda \in \mathbf{C}$, then $f \in W^{n+2,2}_{\text{loc}}(\Omega)$. In particular, if $f \in L^2_{\text{loc}}(\Omega)$ and $-\Delta f = \lambda f$, then f is a smooth function on Ω. We make no comments about the boundary behaviour of f.

Proof Our starting point is the identity

$$-\int_{\mathbf{R}^N} f\Delta(\phi\psi)\mathrm{d}^N x = \lambda\int_{\mathbf{R}^N} f(\phi\psi)\mathrm{d}^N x + \int_{\mathbf{R}^N} g(\phi\psi)\mathrm{d}^N x$$

valid for all $\phi \in C_c^\infty(\Omega)$ and $\psi \in \mathcal{S}$. Suppose inductively that $f \in W^{m,2}_{\text{loc}}(\Omega)$ for some m. Then

$$-\int_{\mathbf{R}^N} f\{\phi\Delta\psi + 2\nabla\phi\cdot\nabla\psi + \psi\Delta\phi\}\mathrm{d}^N x = \lambda\int_{\mathbf{R}^N}(f\phi)\psi\mathrm{d}^N x + \int_{\mathbf{R}^N}(g\phi)\psi\mathrm{d}^N x,$$

where $f\phi$, $f\nabla\phi$ and $f\Delta\phi$ all lie in H^m. Taking Fourier transforms we obtain

$$\int_{\mathbf{R}^N}\left\{(f\phi)\hat{}(y)|y|^2\hat{\psi}(y) + 2i(f\nabla\phi)\hat{}(y)\cdot y\hat{\psi}(y) + (f\Delta\phi)\hat{}(y)\hat{\psi}(y)\right\}\mathrm{d}^N y$$

$$= \lambda\int_{\mathbf{R}^N}(f\phi)\hat{}(y)\hat{\psi}(y)\mathrm{d}^N y + \int_{\mathbf{R}^N}(g\phi)\hat{}(y)\hat{\psi}(y)\mathrm{d}^N y.$$

But $\hat{\psi}$ is an arbitrary element of $\mathcal{S}$, so

$$|y|^2(f\phi)\hat{}(y) = -2iy\cdot(f\nabla\phi)\hat{}(y) - (f\Delta\phi)\hat{}(y) + \lambda(f\phi)\hat{}(y) + (g\phi)\hat{}(y) \quad (6.2.2)$$

for almost every $y \in \mathbf{R}^N$. Writing $k(y)$ in place of the left-hand side and assuming that $m \le n$ we deduce that

$$\int_{\mathbf{R}^N} (1+|y|^2)^{m-1}|k(y)|^2 \mathrm{d}^N y < \infty.$$

Hence

$$\int_{\mathbf{R}^N} (1+|y|^2)^{m+1}|(f\phi)\hat{}(y)|^2 \mathrm{d}^N y < \infty.$$

Induction on m now establishes that $f \in W^{n+1,2}_{\mathrm{loc}}(\Omega)$. This implies in particular that $f\nabla\phi \in H^{n+1}$. Returning to (6.2.2) we now see that

$$\int_{\mathbf{R}^N} (1+|y|^2)^{n}|k(y)|^2 \mathrm{d}^N y < \infty,$$

which implies that $f \in W^{n+2,2}_{\mathrm{loc}}(\Omega)$. If $g = 0$ then we find that $f\phi \in H^n$ for all $\phi \in C_c^\infty(\Omega)$ and all n. This implies that f is smooth in Ω by an application of Theorem 3.7.6. □

Example 6.2.7 It is well known (but not proved in this text) that harmonic functions, defined as solutions of $\Delta f = 0$ in a region, are always C^∞ functions. The same holds for solutions of $Lf = 0$ where L is any second order elliptic operator with smooth coefficients. In this example we construct a simple second order elliptic operator in divergence form and an L-harmonic function which is continuous but not Lipschitz continuous in the interior of the region concerned.

We describe the example only in polar coordinates, and leave the reader to convert to Euclidean coordinates. Define H formally for functions f on $\mathbf{R}^2$ by

$$Hf := -\frac{1}{r}\frac{\partial}{\partial r}\left(r\frac{\partial f}{\partial r}\right) - \frac{c^2}{r^2}\frac{\partial^2 f}{\partial \theta^2},$$

where $0 < c < 1$. The associated quadratic form is defined on $C_c^\infty(\mathbf{R}^2)$ by

$$Q(f) := \int_{\mathbf{R}^2}\left(\left|\frac{\partial f}{\partial r}\right|^2 + \frac{c^2}{r^2}\left|\frac{\partial f}{\partial \theta}\right|^2\right) r\,\mathrm{d}r\,\mathrm{d}\theta$$

and satisfies

$$c^2 Q_0(f) \le Q(f) \le Q_0(f)$$

for all $f \in C_c^\infty(\mathbf{R}^2)$, where

$$Q_0(f) := \int_{\mathbf{R}^2} |\nabla f|^2 \,\mathrm{d}x\,\mathrm{d}y.$$

In Euclidean coordinates H is a uniformly elliptic operator in divergence form whose coefficients are smooth except at 0.

Now put

$$f(r,\theta) := r^c \sin(\theta)$$

for all $r \geq 0$ and $\theta \in [0, 2\pi]$. An easy calculation shows that $f \in W^{1,2}_{\mathrm{loc}}(\mathbf{R}^2)$ and that $Hf(r,\theta) = 0$ in the classical sense for all $r \neq 0$. A slightly more careful calculation shows that $Hf = 0$ in the weak sense on $\mathbf{R}^2$ although f is not Lipschitz continuous at 0. □

6.3 The general case

Still assuming Dirichlet boundary conditions, we now turn to the general uniformly elliptic second order differential operators defined in Section 6.1 with the aid of quadratic form techniques. We first prove a weak generalization of Lemma 6.2.1.

Theorem 6.3.1 *For all bounded regions $\Omega \subseteq \mathbf{R}^N$ and all real measurable coefficients $a_{i,j}(x)$ and $b(x)$ satisfying the conditions (6.1.2) and (6.1.3), the operator H has empty essential spectrum and compact resolvent. The eigenvalues $\{\lambda_n\}_{n=1}^{\infty}$ of H written in increasing order and repeated according to multiplicity satisfy*

$$c_1 n^{2/N} \leq \lambda_n \leq c_2 n^{2/N}$$

for some $c_1, c_2 > 0$ and all $n \geq 1$.

Proof We compare the eigenvalues μ_n of $H_0 := -\Delta$ given by

$$\mu_n(\Omega) = \inf\{\mu(L) : L \subseteq C_c^\infty(\Omega) \text{ and } \dim L = n\},$$

where

$$\mu(L) := \sup\left\{Q_0(f) : f \in L \text{ and } \int_\Omega |f|^2 \leq 1\right\}$$

and Q_0 is defined by (6.2.1), with the eigenvalues λ_n of H given by

$$\lambda_n(\Omega) := \inf\{\lambda(L) : L \subseteq C_c^\infty(\Omega) \text{ and } \dim L = n\}, \tag{6.3.1}$$

where

$$\lambda(L) := \sup\left\{Q(f) : f \in L \text{ and } \int_\Omega b|f|^2 \leq 1\right\} \tag{6.3.2}$$

and Q is defined by (6.1.4). It follows immediately from (6.1.2) and (6.1.3) that

$$c^{-2}\mu_n \le \lambda_n \le c^2\mu_n$$

for all $n \ge 1$. Now the eigenvalues of H_0 depend monotonically upon the region Ω and so can be bounded above (resp. below) by the eigenvalues of cubes which are contained in (resp. contain) Ω. It follows that $\lim_{n\to\infty}\lambda_n = \infty$. The variational theory of Section 4.5 now implies that H has empty essential spectrum and compact resolvent, and that λ_n are its eigenvalues written in increasing order and repeated according to multiplicity. The last statement of the theorem now follows by an application of Corollary 6.2.2. □

Theorem 6.3.2 *The eigenvalues $\{\lambda_n\}_{n=1}^{\infty}$ of H are monotonic decreasing functions of the region Ω, monotonic increasing functions of the coefficient matrix $\{a_{i,j}(x)\}$ and monotonic decreasing functions of the coefficients $b(x)$.*

Proof This is immediate from the equations (6.3.1) and (6.3.2). □

Once again one can ask about stability of the eigenvalues under small perturbations of the coefficients. If by this one means uniformly small perturbations then this follows easily from (6.3.1) and (6.3.2). For more general notions of small perturbation one is once again at the forefront of current research.

We next turn to the behaviour of eigenfunctions of these operators near the boundary of Ω. A complete analysis is more or less impossible because of the complications which may occur. We have already seen that for the Laplacian in two dimensions the rate at which the eigenfunctions decay to zero at a vertex depends upon the angle of that vertex. For more general operators there is the additional complication that the rate of decay near some boundary point may depend upon discontinuities of the coefficients at that boundary point.We illustrate this phenomenon by means of a simple example, which the reader will no doubt be able to elaborate upon.

Example 6.3.3 Let Ω be a bounded region in $\mathbf{R}^2$ with piecewise smooth boundary. Suppose that $0 \in \partial\Omega$ and that the x_2-axis is tangential to $\partial\Omega$ at 0. Let

$$Hf := -\sum_{i,j}\frac{\partial}{\partial x_i}\left\{a_{i,j}(x)\frac{\partial f}{\partial x_j}\right\}.$$

in $L^2(\Omega, \mathrm{d}^2x)$ subject to Dirichlet boundary conditions, where $a_{1,1} = 1+\alpha^2$, $a_{1,2} = a_{2,1} = \mp\alpha$ depending upon whether $\pm x_2 > 0$ and $a_{2,2} = 1$.

We shall show that the eigenfunctions of H do not decay linearly near the origin even if the boundary is smooth there. We first define the map $T : \mathbf{R}^2 \to \mathbf{R}^2$ by

$$T(x_1, x_2) := (x_1 \pm \alpha x_2, x_2),$$

where $\alpha > 0$ and one chooses $\pm$ depending upon the sign of x_2. Let $\tilde{\Omega}$ be the image of Ω by the map T.

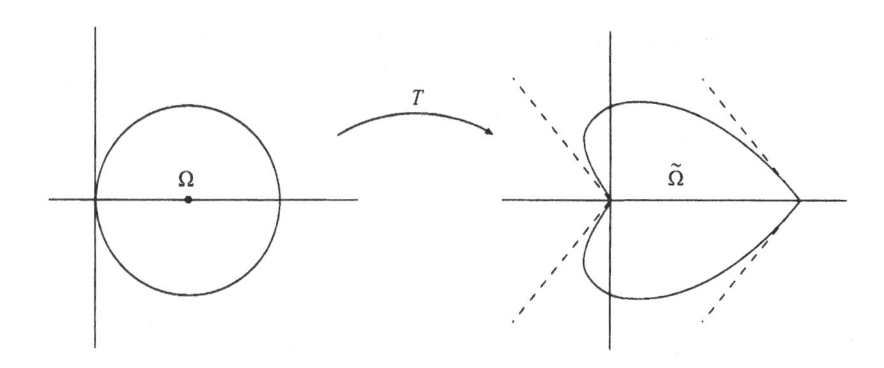

Since T has determinant 1 everywhere, the formula

$$Uf(x_1, x_2) := f(x_1 \pm \alpha x_2, x_2)$$

defines a unitary map U of $L^2(\tilde{\Omega}, \mathrm{d}^2x)$ onto $L^2(\Omega, \mathrm{d}^2x)$. If $f, g \in W_0^{1,2}(\tilde{\Omega})$ and Q is the quadratic form associated with H then a routine computation shows that

$$Q(Uf, Ug) = \int_{\tilde{\Omega}} \langle \nabla(f), \nabla(g) \rangle \mathrm{d}^2x.$$

In other words $U^{-1}HU$ is unitarily equivalent to the operator $H_0 = -\Delta$ acting on $L^2(\tilde{\Omega}, \mathrm{d}^2x)$ subject to Dirichlet boundary conditions. In particular the eigenfunctions of the two operators correspond under U. The rate of decay of the eigenfunctions of the two operators at 0 must therefore be the same. However, the region $\tilde{\Omega}$ has a corner at 0, so the calculations of Example 6.2.5 show that the eigenfunctions of H decay non-linearly at 0.

This proof is not complete because Example 6.2.5 only treated one particular region with a corner, and not the general case. However, we can at least observe that our calculation is valid if Ω is chosen so that $\tilde{\Omega}$ is the sectorial region considered in Example 6.2.5. We have thus established that the rate of decay of an eigenfunction at a boundary point does not depend only upon the geometry near that boundary point if the coefficients of the operator are discontinuous at the point in question. □

Exercises

6.1 Write down the symbol, in the sense of Section 3.5, associated with an operator of the form (6.1.1), assuming that the coefficients $a_{i,j}(x)$ are C^1 functions of x.

6.2 Let H be the operator with domain $C_c^\infty(\Omega)$ associated with the symbol

$$a(x,y) := \sum_{i,j} a_{i,j}(x)y_i y_j + \sum_i b_i(x)y_i + c(x).$$

Find necessary and sufficient conditions on the coefficients, assumed to be C^1, for the operator to be symmetric.

6.3 Let $f \in L^2(\mathbf{R}^N)$ and suppose that there exist $f_r \in L^2(\mathbf{R}^N)$ for $1 \le r \le N$ such that

$$\int_{\mathbf{R}^N} f \frac{\partial \phi}{\partial x^r} \mathrm{d}^N x = -\int_{\mathbf{R}^N} f_r \phi \, \mathrm{d}^N x$$

for all r and all $\phi \in C_c^\infty(\mathbf{R}^N)$. Use the method of proof of Theorem 3.7.3 to show that this identity also holds for all $\phi \in \mathcal{S}$. Deduce that the two definitions of $W^{1,2}(\mathbf{R}^N)$ in Sections 3.7 and 6.1 coincide.

6.4 Let $f \in W_0^{1,2}(\Omega)$ and $g \in C^\infty(\overline{\Omega})$. Prove that $fg \in W_0^{1,2}(\Omega)$ by finding explicit expressions for its weak partial first derivatives.

6.5 Let T be the triangular region $\{(x,y) \in \mathbf{R}^2 : 0 < x < 1, 0 < y < 1, 0 < x < y\}$ and let $H := -\overline{\Delta}$ in $L^2(T)$ subject to Dirichlet boundary conditions. By exploiting known results for the unit square and reflection symmetry about the line $x = y$, find all of the eigenvalues and eigenfunctions of H.

6.6 Let λ be the smallest eigenvalue of $H := -\overline{\Delta}$ subject to Dirichlet boundary conditions in $L^2(B)$, where B is the unit ball in $\mathbf{R}^N$.

Prove that

$$\frac{N\pi^2}{4} \le \lambda \le \frac{N^2\pi^2}{4}$$

by comparing the ball with suitable cubes in $\mathbf{R}^N$.

6.7 Let $\lambda(n)$ denote the nth eigenvalue of $H := -\overline{\Delta}$ subject to Dirichlet boundary conditions in $L^2(S)$ where $S := (0,1) \times (0,1)$. Compute $\lambda(10^6)$ to within an error of 1%. Carry out a similar task for Neumann boundary conditions.

6.8 Let $\Omega := \{re^{i\theta} : 0 < r < 1 + \delta \sin(n\theta)\}$ where $0 < \delta < 1$ and n is a positive integer. Obtain upper and lower bounds on the smallest eigenvalue of $H := -\overline{\Delta}$ acting on $L^2(\Omega)$ subject to Dirichlet boundary conditions, by comparison with the results for appropriate balls.

6.9 Let $H := -\overline{\Delta}$ subject to Dirichlet boundary conditions on the rectangle $(0,1) \times (0,\alpha)$. Prove that if α^2 is irrational then every eigenvalue of H is of multiplicity one.

6.10 Let $H := -\overline{\Delta}$ subject to Dirichlet boundary conditions on the unit square $(0,1)\times(0,1)$. Find the smallest eigenvalue of H which is of multiplicity 3. Are there eigenvalues of multiplicity 4?

6.11 Let $\Omega := U_1 \cup U_2 \cup S$, where

$$\begin{aligned} U_1 &:= \{(x,y) : 0 < x < 1,\ 0 < y < 1\},\\ U_2 &:= \{(x,y) : 0 < x < \delta,\ 1 < y < 1+\delta\},\\ S &:= \{(x,y) : 0 < x < \delta,\ y = 1\} \end{aligned}$$

for some small $\delta > 0$. Obtain upper and lower bounds on the eigenvalues of the operator $H := -\overline{\Delta}$ on $L^2(\Omega)$ subject to Dirichlet boundary conditions by comparing Ω with the regions $U_1 \cup U_2$ and $U_3 := \{(x,y) : 0 < x < 1, 0 < y < 1+\delta\}$.

7

Neumann boundary conditions

7.1 Properties of the $W^{1,2}$ spaces

The analysis of the Laplacian and of more general second order elliptic operators subject to Neumann boundary conditions is intrinsically harder than the corresponding problem for Dirichlet boundary conditions. However, Neumann boundary conditions are directly relevant to a number of physical problems. These include the flow of a fluid through a channel or past an obstacle, the flow of heat in a medium with an insulated boundary, and the vibration of a membrane at those parts of the boundary which are free to move.

The present section is devoted to technical preparations for studying this problem. Our approach will depend heavily upon what is called the $W^{1,2}$ extension property for the region Ω. This is not universally valid, but has been proved under considerably weaker conditions on Ω than those we assume below. Our main result, Theorem 7.1.7, depends upon a series of lemmas, some of which are of interest for their own sake. The first reconciles the definitions of $W^{1,2}$ and of $W_0^{1,2}$ given in Section 6.1 with those of Example 4.4.4.

Lemma 7.1.1 *A function f on the bounded interval $[a,b]$ lies in $W^{1,2}(a,b)$ if and only if there exist a constant c and a function $g \in L^2(a,b)$ such that*

$$f(x) = c + \int_a^x g(s)\mathrm{d}s \tag{7.1.1}$$

for all $x \geq a$. All such functions f are uniformly continuous on $[a,b]$. Moreover, g is equal to the weak derivative of f. A function f of the form (7.1.1) lies in $W_0^{1,2}(a,b)$ if and only if one also has $f(a) = f(b) = 0$, or equivalently $c = 0$ and $\int_a^b g(s)\mathrm{d}s = 0$.

Note Elements of L^2 or $W^{1,2}$ are not actually functions but equivalence classes of functions under the relation of almost everywhere equality. When we say that $f \in W^{1,2}$ is continuous we mean that there is a continuous function in the equivalence class. It is easy to show that there cannot be more than one such continuous function.

Proof If f is of the form (7.1.1) and $\phi \in C_c^\infty(a,b)$ then

$$\begin{aligned}\int_a^b \phi'(s)f(s)\mathrm{d}s &= c\int_a^b \phi'(s)\mathrm{d}s + \iint_{a\le s\le x\le b} \phi'(x)g(s)\mathrm{d}s\mathrm{d}x\\ &= \int_a^b (\phi(b)-\phi(s))g(s)\mathrm{d}s\\ &= -\int_a^b \phi(s)g(s)\mathrm{d}s.\end{aligned}$$

Therefore $f \in W^{1,2}(a,b)$ and $f' = g$ in the weak sense.

Conversely let $f \in W^{1,2}(a,b)$ and $f' = g$. If we define h by

$$h(x) := \int_a^x g(s)\mathrm{d}s,$$

then $f - h \in W^{1,2}(a,b)$ and $(f-h)' = 0$; in other words $\langle f-h, \phi'\rangle = 0$ for all $\phi \in C_c^\infty(a,b)$. The set of such ϕ' is norm dense in $\{\psi \in L^2(a,b) : \langle \psi, 1\rangle = 0\}$. Therefore $f - h = c1$ for some constant c, and f is of the form (7.1.1).

We now consider the question of whether such a function lies in $W_0^{1,2}(a,b)$. An elementary calculation shows that $f \in C_c^\infty(a,b)$ if and only if it is of the form (7.1.1) with $c = 0$, $g \in C_c^\infty(a,b)$ and $\int_a^b g(s)\mathrm{d}s = 0$. A function f of the form (7.1.1) lies in the $W^{1,2}(a,b)$-closure of this subspace if and only if $c = 0$ and g is an L^2 norm limit of a sequence $g_n \in C_c^\infty(a,b)$ for which $\int_a^b g_n(s)\mathrm{d}s = 0$. This set is as described. □

Lemma 7.1.2 *Let S be a region in $\mathbf{R}^{N-1}$ and let $0 < c < \infty$. Let*

$$V := \{w = (u,v) \in \mathbf{R}^N : u \in S \text{ and } -c < v < c\},$$
$$V^- := \{w = (u,v) \in \mathbf{R}^N : u \in S \text{ and } -c < v < 0\}.$$

Then the formula

$$Ef(w) := f((u, -|v|))$$

defines a bounded linear extension operator from $W^{1,2}(V^-)$ to $W^{1,2}(V)$.

Proof It is clear that $Ef \in L^2(V)$, and that $\partial(Ef)/\partial u_i \in L^2(V)$ for all

$1 \le i \le N-1$. Since $\partial f/\partial v \in L^2(V^-)$ we see that $\partial f(u,v)/\partial v \in L^2(-c,0)$ for almost every $u \in S$. For such u, $f(u,\cdot)$ is a continuous function on $[-c,0]$ and is given by the formula (7.1.1). The extension $Ef(u,\cdot)$ of $f(u,\cdot)$ is also continuous on $[-c,c]$ and is given by the formula (7.1.1), where $g(u,-v) = -g(u,v)$ for almost every $u \in S$ and $v \in [-c,c]$. A second application of Lemma 7.1.1 now proves that $\partial(Ef)/\partial v \in L^2(V)$. We conclude that $Ef \in W^{1,2}(V)$ and that

$$\|Ef\|_{W^{1,2}(V)} = 2^{1/2}\|f\|_{W^{1,2}(V^-)}. \qquad \square$$

In our next two lemmas we show that the definition of $W^{1,2}(\Omega)$ is invariant under smooth diffeomorphisms of the region. Let $\mathscr{B}$ denote the set of all sequences $b := (b_0, ..., b_N)$ of functions $b_i \in \mathbf{C}^\infty(\overline{\Omega})$. To each such sequence we define a first order differential operator on $C_c^\infty(\Omega)$ by means of the formula

$$L_b f(x) := b_0(x)f(x) + \sum_{i=1}^{N} b_i(x)\frac{\partial f}{\partial x_i}. \qquad (7.1.2)$$

More precisely we say that $f \in L^2(\Omega)$ lies in the domain of the operator L_b if there exists $g \in L^2(\Omega)$ such that

$$\int_\Omega f\left\{b_0\phi - \sum_{i=1}^{N}\frac{\partial(b_i\phi)}{\partial x_i}\right\}\mathrm{d}^N x = \int_\Omega g\phi\,\mathrm{d}^N x \qquad (7.1.3)$$

for all $\phi \in C_c^\infty(\Omega)$. We then write $L_b f := g$.

Lemma 7.1.3 *A function $f \in L^2(\Omega)$ lies in $W^{1,2}(\Omega)$ if and only if $f \in$ Dom(L_b) for all $b \in \mathscr{B}$.*

Proof If $f \in \mathrm{Dom}(L_b)$ for all $b \in \mathscr{B}$, then one sees that $f \in W^{1,2}(\Omega)$ by applying (7.1.3) to the unit vector fields in the directions of the coordinate axes. The converse is an application of (7.1.2), interpreted in the weak sense. $\square$

Our next lemma is of great importance for the extension of the ideas in this book to differential manifolds. It implies for example that there is a canonical definition of $W^{1,2}(M)$ for any compact differential manifold M, independent of any particular choice of covering by coordinate patches. Our use of it, although important, is much more mundane.

Lemma 7.1.4 *Let Ω and Ω' be two bounded regions in $\mathbf{R}^N$ for which there exists a smooth diffeomorphism τ taking $\overline{\Omega}$ onto $\overline{\Omega'}$. If we define $(Tf)(x) :=$*

$f(\tau^{-1}x)$ then T is a bounded invertible linear operator taking $W^{1,2}(\Omega)$ onto $W^{1,2}(\Omega')$, and $W_0^{1,2}(\Omega)$ onto $W_0^{1,2}(\Omega')$.

Proof Lemma 7.1.3 gives a characterisation of $W^{1,2}$ in terms of a class of differential operators which is invariant as a class with respect to diffeomorphisms of the region and with respect to the multiplication of Lebesgue measure by a positive smooth density. The second point is important because the Jacobean of the diffeomorphism is positive and smooth on $\overline{\Omega}$. The boundedness of T is proved by use of the closed graph theorem. Since $T\big(C_c^\infty(\Omega)\big) = C_c^\infty(\Omega')$ the last statement of the lemma follows by taking limits in the $W^{1,2}$ norms. □

We now return to the study of the extension property. We assume below that the region Ω is bounded and has a smooth boundary. We describe a particular way of covering the boundary by coordinate patches, which will be used in the analysis of both Dirichlet and Neumann boundary conditions.

Construction 7.1.5 Let $a(x) := \{a_{i,j}(x)\}$ be a real symmetric matrix depending smoothly upon $x \in \overline{\Omega}$ and satisfying $c^{-1}1 \le a(x) \le c1$ for some $c \ge 1$ and all $x \in \overline{\Omega}$. For every point $x \in \partial\Omega$ there is a unit outward pointing normal vector $n(x)$ which depends smoothly upon x. If we define $\tilde{n}(x)$ for all $x \in \partial\Omega$ by $\tilde{n}_i(x) := \sum_{j=1}^N a_{i,j}(x)n_j(x)$, then $\tilde{n}(x)\cdot n(x) > 0$, so $\tilde{n}(x)$ is non-tangential. The reason for using $\tilde{n}(x)$ in the constructions below instead of $n(x)$ will appear in Theorem 7.2.1.

Given $a \in \partial\Omega$ there exists a bounded open neighbourhood S of $0 \in \mathbf{R}^{N-1}$ and a smooth non-degenerate map $\sigma : S \to \partial\Omega$ such that $\sigma(0) = a$. An application of the inverse function theorem now implies that the smooth map $\tau : V = S \times (-c, c) \to \mathbf{R}^N$ defined by

$$\tau(u,v) := \sigma(u) + v\,\tilde{n}(\sigma(u))$$

is non-degenerate if S and $c > 0$ are small enough. Moreover, $\tau(u,v) \in \Omega$ if and only if $(u,v) \in V^- := S \times (-c,0)$, and $\tau(u,v) \in \partial\Omega$ if and only if $(u,v) \in S \times \{0\}$. The image $U := \tau(V)$ is a 'coordinate neighbourhood' of a in $\mathbf{R}^N$.

The set of all such coordinate neighbourhoods forms an open covering of $\partial\Omega$, which has a finite subcovering $\{U_i\}_{i=1}^m$ by compactness. In subsequent discussions we indicate the dependence of τ, S, V etc. upon i by a

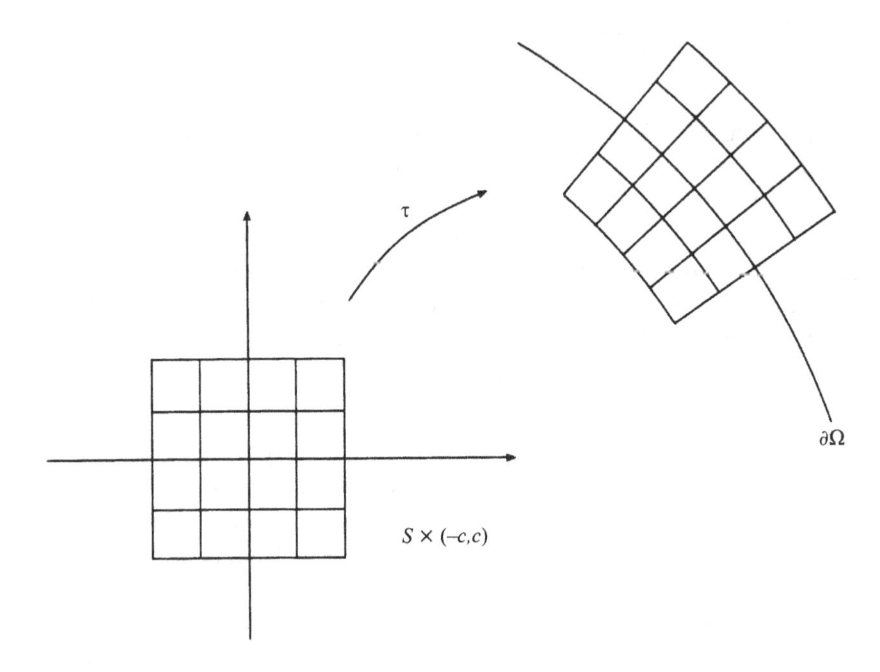

subscript in the obvious manner. There exists an open set U_0 such that $\overline{U_0} \subseteq \Omega$ and $\cup_{i=0}^m U_i$ is a finite open covering of $\overline{\Omega}$. □

Given the covering of $\overline{\Omega}$ constructed above, we will need a related C^∞ partition of the identity. By this we mean a set of non-negative smooth functions $\{\phi_i\}_{i=0}^m$ on $\mathbf{R}^N$ such that $\text{supp}(\phi_i) \subseteq U_i$ for $0 \le i \le m$ and $\sum_{i=0}^m \phi_i(x) = 1$ for all $x \in \overline{\Omega}$.

Lemma 7.1.6 *For any covering of $\overline{\Omega}$ by open subsets $\{U_i\}_{i=0}^m$, a smooth partition of the identity does exist.*

Proof We put $U_{m+1} := \mathbf{R}^N \backslash \overline{\Omega}$ to obtain a covering of the whole of $\mathbf{R}^N$. We then produce a covering of $\mathbf{R}^N$ by slightly smaller sets. We define the non-negative continuous functions a_i on $\mathbf{R}^N$ by $a_i(x) := \min\{|x-y| : y \notin U_i\}$. Since $a(x) := \sum_{i=0}^{m+1} a_i(x) > 0$ for all $x \in \mathbf{R}^N$ and $\lim_{|x|\to\infty} a(x) = +\infty$, there exists $\varepsilon > 0$ such that $a(x) > (m+2)\varepsilon$ for all $x \in \mathbf{R}^N$. Therefore $\mathbf{R}^N$ is also covered by the sets $U_{i,\varepsilon} := \{x : a_i(x) > \varepsilon\} \subset U_i$.

Now let $\{g_i\}_{i=0}^{m+1}$ be smooth non-negative functions on $\mathbf{R}^N$ such that $g_i(x) \ge 1$ if $x \in \overline{U}_{i,2\varepsilon/3}$ and $g_i(x) = 0$ if $x \notin U_{i,\varepsilon/3}$. It is immediate that each g_i has support in U_i and that $g(x) := \sum_{i=0}^{m+1} g_i(x) \ge 1$ on $\mathbf{R}^N$. We now define $\phi_i(x) := g_i(x)/g(x)$ for $0 \le i \le m$ to complete the construction of the partition of the identity. □

Theorem 7.1.7 *If Ω is bounded with a smooth boundary then there exists a bounded linear 'extension' operator $E : W^{1,2}(\Omega) \to W^{1,2}(\mathbf{R}^N)$ such that $(Ef)(x) = f(x)$ for all $f \in W^{1,2}(\Omega)$ and all $x \in \Omega$.*

Proof This proof is easier to understand visually than by equations, and we urge the reader to draw some diagrams to represent the mappings and the supports of the various functions. Let $\{V_i\}_{i=1}^m$, $\{\tau_i\}_{i=1}^m$ and $\{U_i\}_{i=0}^m$ be the objects constructed above. If $1 \le i \le m$, then using Lemma 7.1.4 we define the bounded operator $T_i : W^{1,2}(\Omega) \to W^{1,2}(V_i^-)$ by $(T_i f)(w) := f(\tau_i w)$, where $w \in V_i^-$. We then define the bounded operator $F_i : W^{1,2}(\Omega) \to W^{1,2}(V_i)$ by $F_i(f) := E_i T_i(\phi_i f)$, where E_i is the extension operator from $W^{1,2}(V_i^-)$ to $W^{1,2}(V_i)$ defined in Lemma 7.1.2. Because ϕ_i has compact support in U_i, an examination of the definition of E_i shows that $F_i(f)$ has compact support in V_i. An application of Theorem 6.1.6 now shows that $F_i(f) \in W_0^{1,2}(V_i)$ for all $f \in W^{1,2}(\Omega)$. Using Lemma 7.1.4 once again, we define $G_i : W_0^{1,2}(V_i) \to W^{1,2}(\mathbf{R}^N)$ by

$$(G_i g)(x) := \begin{cases} g(\tau_i^{-1} x) & \text{if } x \in U_i, \\ 0 & \text{otherwise.} \end{cases}$$

We then finally define $E : W^{1,2}(\Omega) \to W^{1,2}(\mathbf{R}^N)$ by

$$(Ef)(x) := \phi_0(x) f(x) + \sum_{i=1}^{m} (G_i F_i f)(x).$$

If $x \in \Omega$, then $(G_i F_i f)(x) = \phi_i(x) f(x)$ for all $i \ge 1$. Therefore

$$(Ef)(x) = \sum_{i=0}^{m} \phi_i(x) f(x) = f(x)$$

for all $x \in \Omega$, and E has all the properties stated in the theorem. □

Corollary 7.1.8 *If Ω is bounded with a smooth boundary, then the space $C^\infty(\overline{\Omega})$ is norm dense in $W^{1,2}(\Omega)$ for the norm $|||\ |||$. Indeed the set of restrictions to $\overline{\Omega}$ of functions in $\mathcal{S}$ has the same property.*

Proof Let $f \in W^{1,2}(\Omega)$ and let $g := Ef \in W^{1,2}(\mathbf{R}^N)$. It follows from Lemma 3.7.1 that the Fourier transform of g satisfies

$$|||g|||^2 = \int_{\mathbf{R}^N} (1 + |y|^2) |\mathcal{F} g(y)|^2 \mathrm{d}^N y < \infty.$$

By Theorem 3.7.3 there exists a sequence $\{g_n\}_{n=1}^{\infty} \subseteq \mathcal{S}$ such that $\lim_{n\to\infty} |||g - g_n||| = 0$. The restrictions of g_n to Ω converge to f in the norm of $W^{1,2}(\Omega)$. □

A region Ω in $\mathbf{R}^N$ for which there exists an operator E with the properties of Theorem 7.1.7 is said to have the extension property. This property holds for many but not all regions in $\mathbf{R}^N$, and in particular holds for all bounded regions with piecewise smooth or even Lipschitz boundaries. The following example shows that it is not always valid.

Example 7.1.9 Let $\alpha > 1$ and consider the region $\Omega_\alpha \subseteq \mathbf{R}^2$ defined by

$$\Omega_\alpha := \{(x, y) : 0 < x < 1 \text{ and } |y| < x^\alpha\}.$$

Consider also the function $f : \Omega_\alpha \to \mathbf{R}$ defined by $f(x, y) := x^{-\beta}$ where $\beta > 0$. We have

$$|||f|||^2 = c_1 \int_0^1 x^{-2\beta+\alpha} \mathrm{d}x + c_2 \int_0^1 x^{-2\beta-2+\alpha} \mathrm{d}x$$

so $f \in W^{1,2}(\Omega_\alpha)$ if and only if $\beta < (\alpha - 1)/2$. A similar calculation shows that $f \in L^p(\Omega_\alpha)$ if and only if $p < (\alpha + 1)/\beta$. If $0 < \beta < (\alpha - 1)/2$ and f could be extended to a function $g \in W^{1,2}(\mathbf{R}^2)$, then it would be immediate from Theorem 3.6.4 that $g \in L^p(\mathbf{R}^2)$ for all $2 \le p < \infty$. The contradiction implies that Ω_α does not have the extension property if $\alpha > 1$.

More generally one cannot expect any bounded region to have the extension property if it possesses an outwardly pointing cusp. The precise conditions can be determined by modifications of the above calculations. □

7.2 Neumann boundary conditions

The study of Neumann boundary conditions differs from the study of Dirichlet boundary conditions in several important respects. The first is that one has no monotonicity properties of the eigenvalues as the region expands or contracts. The second is that the natural general definition of Neumann boundary conditions for bounded regions has the property that if the boundary is sufficiently peculiar the Neumann Laplacian may have non-compact resolvent. Research completed in the last five years establishes that one may change the spectrum as much as one likes by means of a perturbation of the boundary in as small a neighbourhood of one point of the boundary as one likes! Fortunately it is possible to avoid these pathologies by imposing regularity conditions on the boundaries of the regions considered.

If Ω is a bounded region in $\mathbf{R}^N$ with a smooth boundary we consider the operator H on $L^2(\Omega, b(x)\,\mathrm{d}^N x)$ defined formally by

$$Hf := -b(x)^{-1} \sum_{i,j=1}^{N} \frac{\partial}{\partial x_i} \left\{ a_{i,j}(x) \frac{\partial f}{\partial x_j} \right\}. \qquad (7.2.1)$$

We assume that the real coefficients $a_{i,j}(x)$ and $b(x)$ are smooth on $\overline{\Omega}$ and satisfy the conditions

$$c^{-1}1 \le \{a_{i,j}(x)\} \le c\,1 \qquad (7.2.2)$$

and

$$c^{-1} \le b(x) \le c \qquad (7.2.3)$$

for some constant $c \ge 1$ and all $x \in \Omega$. At the end of the section we describe how to relax the conditions on $\partial\Omega$ and on the coefficients of H.

We say that $H_N := H$ satisfies Neumann boundary conditions if we take its initial domain to be the space $\mathcal{D}$ of all smooth functions $f \in C^\infty(\overline{\Omega})$ such that

$$\tilde{n}(x) \cdot \nabla f(x) = \sum_{i,j} a_{i,j}(x) \frac{\partial f}{\partial x_j} n_i(x) = 0 \qquad (7.2.4)$$

for all $x \in \partial\Omega$, where $n(x)$ is the unit normal vector at the point $x \in \partial\Omega$ and $\tilde{n}(x)$ was defined in Construction 7.1.5. If $a_{i,j}(x) = \alpha(x)\delta_{i,j}$, then this reduces to the condition that the normal derivative of f vanishes on $\partial\Omega$. It is an elementary fact that the operator H_N is symmetric on the domain $\mathcal{D}$ subject to Neumann boundary conditions.

The associated quadratic form Q on $\mathcal{D}$ is given by

$$\begin{aligned} Q(f) &:= \langle Hf, f \rangle \\ &= \int_\Omega \sum_{i,j=1}^{N} a_{i,j}(x) \frac{\partial f}{\partial x_i} \frac{\partial \overline{f}}{\partial x_j} \mathrm{d}^N x. \end{aligned} \qquad (7.2.5)$$

Since Q is non-negative H_N has a Friedrichs extension, which we denote by the same symbol. The theorem below is remarkable in that it shows that the boundary condition (7.2.4) disappears as soon as one passes from the operator H_N to the closure of its quadratic form.

Theorem 7.2.1 *The quadratic form Q is closable on $\mathcal{D}$. Its closure $\overline{Q}$ has domain $W^{1,2}(\Omega)$ and is still given by the formula (7.2.5).*

Proof By Corollary 7.1.8 we need only show that for every $f \in C^\infty(\overline{\Omega})$ there exists a sequence $f_n \in \mathcal{D}$ such that $\lim_{n\to\infty} |||f_n - f||| = 0$. By applying the partition of the identity of Lemma 7.1.6, we need only consider the case where f has support in a single coordinate patch U_i. By using the diffeomorphism of Construction 7.1.5 to transfer to V_i we end up with the following claim, in which we suppress the subscript.

For every $g \in C_c^\infty(S \times (-c, 0])$ there exists a sequence $g_n \in C_c^\infty(S \times (-c, 0])$ such that $\partial g_n / \partial v = 0$ on $S \times \{0\}$ and $\lim_{n\to\infty} |||g_n - g||| = 0$ in $W^{1,2}(S \times (-c, 0))$.

We prove this claim by defining $\{g_n\}_{n=1}^\infty$ as follows. Let $h_n : \mathbf{R} \to \mathbf{R}$ be smooth functions such that $h_n(s) = s$ if $s \le -2/n$, $h_n(s) = 0$ if $s \ge -1/n$ and $0 \le h_n'(s) \le 3$ for all $s \in \mathbf{R}$. If we put $g_n(u, v) := g(u, h_n(v))$ for all $(u, v) \in S \times (-c, 0)$, then $\lim_{n\to\infty} g_n(u, v) = g(u, v)$ for all $(u, v) \in S \times (-c, 0)$. Similar results hold for all of the first order partial derivatives of g_n. We deduce that $\lim_{n\to\infty} |||g_n - g||| = 0$ by applying the dominated convergence theorem. □

The above theorem justifies our saying that the self-adjoint Friedrichs extension H_N satisfies Neumann boundary conditions. It is the case that H_N is essentially self-adjoint on $\mathcal{D}$ under the above conditions, but we will not use this fact because it is not valid for less regular boundaries or coefficients.

Theorem 7.2.2 *If Ω is a bounded region with a smooth boundary, then the Friedrichs extension H_N of the operator defined on the domain $\mathcal{D}$ by (7.2.1) has a compact resolvent. If $\{\mu_n\}_{n=0}^\infty$ (resp. $\{\lambda_n\}_{n=1}^\infty$) are the eigenvalues of H_N (resp. the operator H_D satisfying Dirichlet boundary conditions), then*

$$0 \le \mu_{n-1} \le \lambda_n$$

for all $n \ge 1$.

Proof The compactness of the resolvent of H_N is equivalent to the compactness of the canonical injection i of $W^{1,2}(\Omega)$ into $L^2(\Omega)$ by Exercise 4.2. Let B be a cubical box containing Ω and let K_N be the Neumann Laplacian in $L^2(B)$. This operator has compact resolvent by an obvious modification of the argument of Lemma 6.2.1, so the canonical injection k of $W^{1,2}(B)$ into $L^2(B)$ is compact. By putting $i = RkE$, where $E : W^{1,2}(\Omega) \to W^{1,2}(B)$ is the extension operator of Theorem 7.1.7 and $R : L^2(B) \to L^2(\Omega)$ is the restriction operator, we see that i is compact.

The statement of the theorem comparing the Neumann and Dirichlet eigenvalues is a direct corollary of the variational theory of Section 4.5 and the obvious inclusion $W_0^{1,2}(\Omega) \subseteq W^{1,2}(\Omega)$. □

We finally make some comments about the weakening of the regularity conditions imposed in the theorems of this section. If Ω is any region in $\mathbf{R}^N$ and Q is the quadratic form defined by the integral (7.2.5), where $a(x)$ are measurable uniformly elliptic coefficients, then Q is closed on the domain $W^{1,2}(\Omega)$. The associated self-adjoint operator H is said to satisfy Neumann boundary conditions, even though it is not possible to identify its domain $\mathscr{D}$ in general, and even though a normal direction may not be definable at any point of the boundary $\partial\Omega$. If Ω has the extension property then the conclusions and proof of Theorem 7.2.2 are all still valid.

Another modification of these theorems allows us to treat mixed Neumann–Dirichlet boundary conditions. Let Ω be a bounded region with the extension property, and let S be a closed subset of $\partial\Omega$. We describe how to define the elliptic operator H_S given by the above formulae but subject to Dirichlet boundary conditions on S and to Neumann boundary conditions on $\partial\Omega\backslash S$. We need only assume that the coefficients are measurable.

The only modification needed in our previous theory is to specify the domain of the quadratic form. We take this to be the closure in $W^{1,2}(\Omega)$ of the set of all smooth functions on $\overline{\Omega}$ which vanish in a neighbourhood of S. One obtains the Dirichlet operator by putting $S = \partial\Omega$ and the Neumann operator by putting $S = \emptyset$.

Theorem 7.2.3 *If Ω is a bounded region with the extension property, then H_S has compact resolvent. The eigenvalues $\{\lambda_n(S)\}_{n=1}^{\infty}$ of H_S written in increasing order and repeated according to multiplicity are monotonic increasing functions of the set S.*

Proof Given S, let $W_S^{1,2}$ denote the closed subspace of $W^{1,2}(\Omega)$ defined above. If $S_1 \subseteq S_2$, then $W_0^{1,2}(\Omega) \subseteq W_{S_2}^{1,2} \subseteq W_{S_1}^{1,2} \subseteq W^{1,2}(\Omega)$. The variational formulae of Section 4.5 imply that $0 \le \mu_{n-1} \le \lambda_n(S_1) \le \lambda_n(S_2) \le \lambda_n$ for all $n \ge 1$, in an obvious notation. Therefore $\lim_{n\to\infty} \lambda_n(S_i) = \infty$ and H_{S_i} have compact resolvents. □

At the present time there are no general quantitative results controlling the rate of variation of the eigenvalues under small perturbations of the set S, although mere continuity of the eigenvalues as the sets S_n decrease to the limit set S is easy to establish.

7.3 Computation of eigenvalues

We consider how to find good numerical approximations to the eigenvalues of the operators considered in the last section. Under the conditions of Theorem 7.2.1 it is actually the case that all of the eigenfunctions lie in $\mathscr{D}$. However, this fact is not particularly useful for the calculation of the eigenvalues.

Many numerical methods of computing the eigenvalues of such operators are based upon applying the variational formulae to a dense subspace of $W^{1,2}(\Omega)$. The subspace $\mathscr{D}$ defined above is one candidate, but so is the larger and simpler subspace $C^\infty(\overline{\Omega})$.

The following theorem yields a subspace which is sometimes suitable for numerical computations.

Theorem 7.3.1 *Let P_n denote the space of restrictions to Ω of polynomials of degree at most n. Then P_n is finite-dimensional, and the union $\cup_{n=1}^{\infty} P_n$ is dense in $W^{1,2}(\Omega)$.*

Proof Let $U \subseteq \mathbf{R}^N$ be an open bounded set such that $\overline{\Omega} \subseteq U$ and let $f \in \mathscr{S}$. By the Stone–Weierstrass theorem there exists a sequence $\{p_n\}_{n=1}^{\infty}$ of polynomials which converge uniformly to f on $\overline{U}$. Now let k_s be the mollifiers constructed in Section 3.2. It is easy to show that $p_{n,s} := k_s * p_n$ are polynomials and that $D^\alpha p_{n,s}$ converge uniformly to $D^\alpha(k_s * f)$ on $\overline{\Omega}$ as $n \to \infty$ for all α and all $s > 0$. Therefore $\lim_{n\to\infty} |||p_{n,s} - k_s * f||| = 0$ in $W^{1,2}(\Omega)$. By using Fourier transforms we also see that $\lim_{s\to 0} |||k_s * f - f||| = 0$ in $W^{1,2}(\mathbf{R}^N)$. Therefore $\lim |||p_{n,s} - f||| = 0$ in $W^{1,2}(\Omega)$. The proof is completed by an application of Corollary 7.1.8. □

By restricting the quadratic form Q and the inner product to P_n we obtain a finite matrix eigenvalue problem. The eigenvalues obtained by solving this problem are larger than the eigenvalues of the differential operator, but converge to them as $n \to \infty$.

There is nothing special about the use of polynomials in the above discussion. If one chooses an approriate n-dimensional space of analytic functions on Ω, one can often obtain very accurate approximations to the eigenvalues with surpringly small values of n, for example $n < 10$. This, however, depends upon Ω having a simple geometric structure.

There is another method of computing eigenvalues called the finite element method, which also leads to eigenvalue problems for large symmetric matrices. However, in this case the matrices are sparse (most of the entries are zero), which allows the numerical calculations to be much faster (or alternatively allows the calculations to be performed for much larger matrices). This is not the place to give a detailed description of the finite element method. There is an enormous literature on the subject, as well as many highly refined computer programs. What we can do here is to give the reader an idea of the method.

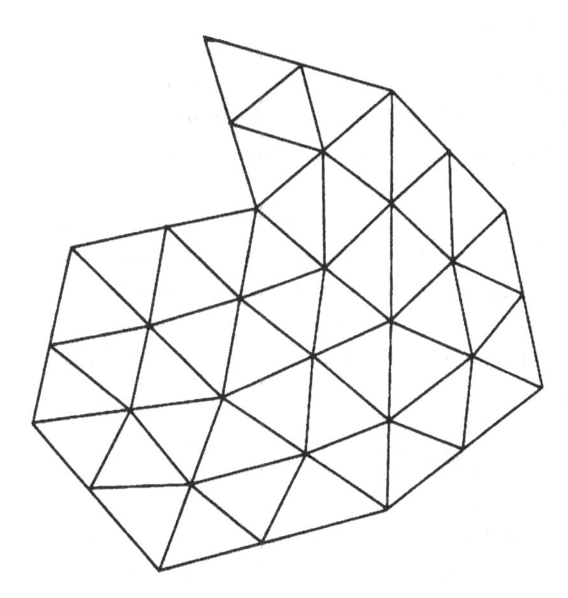

Consider the case where the region Ω is a polygon in $\mathbf{R}^2$. The idea is to subdivide Ω into a large number of small polygons (elements), typically triangles whose angles should not be too small or large, so that their areas are of the same order of magnitude as the square of any of their edges.

The simplest version of the method involves subdividing Ω into elements whose sizes are all of the same order of magnitude. More refined versions reduce the size of the elements near re-entrant corners of the region in a manner expected to increase the overall accuracy of the method; this adjustment is done automatically by the program as the computation proceeds. We then apply the variational method to a linear subspace of $W^{1,2}(\Omega)$ consisting of continuous functions on Ω whose restrictions to each element are polynomials of some specified degree.

The finite element method is particularly simple for second order elliptic operators subject to Neumann boundary conditions, since the functions on each element may then be taken to be affine. Suppose that the region Ω is covered by a finite number of non-overlapping triangles $\{T_s\}_{s\in S}$ whose vertices are all taken from the set $\{x_j\}_{j=1}^n$ of points in $\overline{\Omega}$. Then given any sequence $\{f_j\}_{j=1}^n$ of complex numbers, we can define a function f on Ω as follows. If the triangle T_s has vertices x_i, x_j, x_k, then we define f on T_s to be the unique affine function which equals f_i, f_j, f_k at the stated vertices. It is easy to see that these definitions are compatible at the edges joining two triangles, and this implies that the function f is continuous on Ω.

Theorem 7.3.2 *For such a triangulation of Ω the space L of piecewise affine functions defined above is an n-dimensional subspace of $W^{1,2}(\Omega)$. The variational problem restricted to this subspace only involves sparse matrices.*

Proof Since there is a one-one correspondence between sequences and functions $f \in L$, the space L is n-dimensional. Given i, j, we write $i \sim j$ if $i = j$ or if x_i and x_j are vertices of a common triangle T_s; if the triangles are all very small this implies that x_i and x_j are very close to each other. If $f \in L$, then

$$\|f\|_2^2 = \sum_{s\in S} \int_{T_s} |f|^2 \mathrm{d}^2x = \sum_{i,j=1}^{n} b_{i,j} f_i \overline{f_j}.$$

Each of the integrals in this identity involves only terms f_i, f_j relating to the vertices of a single triangle. Therefore $b_{i,j} = 0$ unless $i \sim j$. This is what we mean by saying that the matrix $\{b_{i,j}\}$ is sparse. Exactly similar considerations apply to the matrix $\{c_{i,j}\}$ defined by

$$Q(f) = \sum_{s\in S} \int_{T_s} \sum_{i,j=1}^{2} a_{i,j}(x) \frac{\partial f}{\partial x_i} \frac{\partial \overline{f}}{\partial x_j} \mathrm{d}^2x = \sum_{i,j=1}^{n} c_{i,j} f_i \overline{f_j}. \qquad \square$$

We mention briefly two variations of the above ideas. If the region is not a polygon (or a polyhedron in higher dimensions), then one makes a further approximation before applying the finite element method. This could be either replacing the boundary by an approximating polygon or using elements which have curved edges at the boundary. In both cases one needs a theorem to the effect that the eigenvalues of the modified problem are close to the true eigenvalues, with error estimates if possible. Secondly if one is interested in Dirichlet boundary conditions, one can simply select polynomials on the boundary elements which vanish in the appropriate sense; Theorem 6.1.5 shows that one then obtains a subspace of $W_0^{1,2}(\Omega)$. One has to solve the same variational problem, but on a certain subspace of the original space. The variational method shows that the Dirichlet eigenvalues are larger than the Neumann eigenvalues for the finite element approximation, as they are for the true problem.

Exercises

7.1 Define the functions f_i on the unit ball $B := \{x \in \mathbf{R}^N : |x| < 1\}$ by $f_i(x) := x_i/|x|$. Prove that $f_i \in W^{1,2}(B)$.

7.2 Let $\{\lambda_n\}_{n=1}^{\infty}$ be the eigenvalues of $H := -\overline{\Delta}$ on $L^2(\Omega)$ subject to either Neumann or Dirichlet boundary conditions. Determine the corresponding eigenvalues for the region $\alpha\Omega := \{\alpha x : x \in \Omega\}$ for any $\alpha > 0$.

7.3 Let $\Omega \subseteq \mathbf{R}^2$ be a bounded region with piecewise smooth boundary, and such that the angle θ at each vertex satisfies $0 < \theta < 2\pi$. By adapting the constructions of Section 7.1, prove that Ω has the extension property.

7.4 Let U be a region in $\mathbf{R}^{N-1}$ and let $-\infty < a < b < \infty$. Prove that for every f in $W^{1,2}\big((a,b) \times U\big)$ there exists a norm continuous function $g : [a,b] \to L^2(U)$ such that $g_s(u) = f(s,u)$ almost everywhere in $(a,b) \times U$. Hint: Start from the formula

$$g_s(u) := f(a,u) + \int_a^s \frac{\partial f}{\partial r}(r,u)\mathrm{d}r,$$

where $a \leq s \leq b$ and $u \in U$.

7.5 Let $\Omega \subseteq \mathbf{R}^2$ be the region

$$\Omega := (-1,1)^2 \setminus \big([0,1) \times \{0\}\big).$$

Use the last exercise to construct a piecewise linear function on

Ω such that $f \in W^{1,2}(\Omega)$ but $f \notin W^{1,2}\left((-1,1)^2\right)$. Deduce that Ω does not have the extension property.

7.6 Let Ω and Ω' be two bounded regions in $\mathbf{R}^N$ and let τ be a smooth diffeomorphism taking $\overline{\Omega}$ one-one onto $\overline{\Omega'}$. Prove that if the Neumann Laplacian on $L^2(\Omega)$ has compact resolvent then the Neumann Laplacian on $L^2(\Omega')$ has compact resolvent.

7.7 Let H_N be the Friedrichs extension of $-\Delta$ acting on $L^2(\Omega)$ subject to Neumann boundary conditions. Assume that

$$\Omega := \{(x,y) : 0 < x < 1 \text{ and } 0 < y < \gamma(x)\} \subseteq \mathbf{R}^2,$$

where $\gamma : [0,1] \to [1,2]$ is smooth with $|\gamma'(x)| \le c$ for all $x \in [0,1]$. By constructing a suitable diffeomorphism $\tau : [0,1]^2 \to \overline{\Omega}$ and using the last exercise, obtain upper and lower bounds on the eigenvalues of H_N in terms of the constant c.

8
Schrödinger operators

8.1 Introduction

In this chapter we turn to the study of Schrödinger operators acting on $L^2(\mathbf{R}^N)$. These are operators such as

$$Hf(x) := -\Delta f(x) + V(x)f(x), \tag{8.1.1}$$

where the real-valued function V on $\mathbf{R}^N$ is called a potential. The symbol of this operator is $a(x, y) := |y|^2 + V(x)$. Much of the impetus for the study of Schrödinger operators comes from quantum theory. In this context a function $f \in L^2(\mathbf{R}^N)$ with $\|f\|_2 = 1$ is called a wave packet or state, and represents the instantaneous configuration of a collection of electrons, atoms and molecules. The operator H is also called the Hamiltonian for historical reasons – quantum theory can be regarded as a non- commutative version of classical Hamiltonian mechanics. The evolution of a quantum system is controlled by the Schrödinger equation

$$\frac{\partial f}{\partial t} = -iHf$$

with solution $f(x,t) = e^{-iHt}f(x,0)$. The total energy $\langle Hf, f\rangle$ of the system is divided between the kinetic energy $\langle -\Delta f, f\rangle$ and the potential energy $\langle Vf, f\rangle$. The smallest eigenvalue of H, if there is one, is called the ground state energy, and the corresponding eigenfunction, called the ground state, is the configuration of the system with the smallest total energy. Other eigenvalues correspond to discrete (quantized) excitations of the system.

The above account is over-simplified in that it does not mention spin or statistics, and assumes that all particles have mass $\frac{1}{2}$. However, it does indicate how the spectral theory of differential operators is of relevance to quantum theory, which is surely the most highly confirmed and important scientific theory of this century.

Our goal in this chapter is to find conditions on the potential V under which one may associate a self-adjoint operator H with the equation (8.1.1). We then investigate the spectral properties of these operators. We cannot possibly match the quantity or the level of sophistication of recent research literature in this book, and select our material on the basis of its technical simplicity and illustrative value. Much more comprehensive accounts of the subject may be found in the volumes of Simon *et al* (1975, 1978, 1987).

Before commencing the detailed theory, we need to mention a mild generalization of the theory of quadratic forms. We say that a self-adjoint operator H on a Hilbert space is semibounded or bounded below if there exists a constant $c \in \mathbf{R}$ such that $\mathrm{Spec}(H) \subseteq [c, \infty)$ or equivalently $\langle Hf, f \rangle \geq c\|f\|^2$ for all $f \in \mathrm{Dom}(H)$; see Theorem 4.3.1. The quadratic form Q_c defined on $\mathrm{Dom}(H)$ by

$$Q_c(f) := \langle Hf, f \rangle - c\|f\|^2$$

is then non-negative, and its closure is associated with the Friedrichs extension of $H - c1$. There is no need to write down explicitly a theory of semibounded operators and forms, since every result in that theory is obtained simply by adding a suitable constant and then quoting the appropriate result for non-negative forms from Section 4.4.

8.2 Definition of the operators

We follow our standard procedure of defining the operators by quadratic form techniques, and do not attempt to discover whether they are essentially self-adjoint on such domains as $C_c^\infty := C_c^\infty(\mathbf{R}^N)$.

We say that a potential $V : \mathbf{R}^N \to \mathbf{R}$ is locally L^p, symbolically $V \in L^p_{\mathrm{loc}}$, if its restriction to every bounded subset of $\mathbf{R}^N$ lies in L^p. We start by considering Schrödinger operators for which the potentials are non-negative and locally L^1. Let Q_0 be the quadratic form associated to $H_0 := -\overline{\Delta}$, as discussed in Example 4.4.7. The domain of Q_0 is $W^{1,2} := W^{1,2}(\mathbf{R}^N)$ and for f in that domain we have

$$Q_0(f) = \int_{\mathbf{R}^N} |\nabla f|^2 \mathrm{d}^N x.$$

Let Q_1 be the quadratic form associated with the multiplication operator $0 \leq V \in L^1_{\text{loc}}$, its domain being

$$\left\{ f : \int_{\mathbf{R}^N} \{1 + V(x)\} |f(x)|^2 \mathrm{d}^N x < \infty \right\}.$$

The proof of our next theorem is longer than it might be, because of a decision to treat rather general potentials. For bounded potentials the proof is essentially a demonstration that $W_0^{1,2}(\mathbf{R}^N) = W^{1,2}(\mathbf{R}^N)$.

Theorem 8.2.1 *If $0 \leq V \in L^1_{\text{loc}}$, then the quadratic form*

$$\begin{aligned} Q(f) &:= Q_0(f) + Q_1(f) \\ &= \int_{\mathbf{R}^N} \{|\nabla f|^2 + V|f|^2\} \mathrm{d}^N x \end{aligned}$$

defined on

$$\mathrm{Dom}(Q) := \mathrm{Dom}(Q_0) \cap \mathrm{Dom}(Q_1)$$

is the form of a non-negative self-adjoint operator H. The space C_c^∞ is a core for Q.

Proof It follows from Theorem 4.4.2 that Q_0 and Q_1 are both lower semicontinuous, and this implies at once that Q is also lower semicontinuous. Since C_c^∞ is clearly contained in Dom(Q), Q is densely defined, and a second application of Theorem 4.4.2 establishes that it is the form of a non-negative self-adjoint operator. We prove that C_c^∞ is a core for Q by a direct if somewhat lengthy approximation procedure.

Part 1 Let $f \in \mathrm{Dom}(Q)$ and let $0 \leq \phi \in C_c^\infty$ satisfy $\phi(x) = 1$ if $|x| < 1$ and $\phi(x) = 0$ if $|x| > 2$. Then put $f_n(x) := f(x)\phi(x/n)$. Elementary calculations establish that f_n lies in both Dom(Q_0) and Dom(Q_1). Moreover,

$$\begin{aligned} &Q(f - f_n) \\ &= \int_{\mathbf{R}^N} |\nabla f(x)\{1 - \phi(x/n)\} - n^{-1} f(x)(\nabla\phi)(x/n)|^2 \mathrm{d}^N x \\ &\quad + \int_{\mathbf{R}^N} V(x)|f(x)|^2 |1 - \phi(x/n)|^2 \mathrm{d}^N x \\ &\leq 2\int_{\mathbf{R}^N} |\nabla f(x)|^2 |1 - \phi(x/n)|^2 \mathrm{d}^N x + 2n^{-2} \int_{\mathbf{R}^N} |f(x)|^2 |(\nabla\phi)(x/n)|^2 \mathrm{d}^N x \\ &\quad + \int_{\mathbf{R}^N} V(x)|f(x)|^2 |1 - \phi(x/n)|^2 \mathrm{d}^N x. \end{aligned}$$

This converges to zero as $n \to \infty$ by the dominated convergence theorem. We conclude that the set of functions $f \in \mathrm{Dom}(Q)$ which are of compact support is a core for Q.

Part 2 Let $f \in \mathrm{Dom}(Q)$ have compact support. Let $n > 0$ and let $F_n : \mathbf{R} \to \mathbf{R}$ be a smooth increasing function such that

$$F_n(s) := \begin{cases} s & \text{if } -n \le s \le n, \\ n+1 & \text{if } s \ge n+2, \\ -n-1 & \text{if } s \le -n-2 \end{cases}$$

and $0 \le F_n'(s) \le 1$ for all $s \in \mathbf{R}$. If we put $f_n(x) := F_n(f(x))$ then $|f_n(x)| \le |f(x)|$ and $\lim_{n\to\infty} f_n(x) = f(x)$ for all $x \in \mathbf{R}^N$. It is immediate by the dominated convergence theorem that

$$\lim_{n\to\infty} Q_1(f - f_n) = \lim_{n\to\infty} \int_{\mathbf{R}^N} V|f - f_n|^2 \mathrm{d}^N x = 0.$$

Also

$$\begin{aligned} \lim_{n\to\infty} Q_0(f - f_n) &= \lim_{n\to\infty} \int_{\mathbf{R}^N} |\nabla f - F_n'(f(x))\nabla f|^2 \mathrm{d}^N x \\ &= \lim_{n\to\infty} \int_{\mathbf{R}^N} |1 - F_n'(f(x))|^2 |\nabla f(x)|^2 \mathrm{d}^N x \\ &= 0. \end{aligned}$$

Therefore the set of bounded functions of compact support in $\mathrm{Dom}(Q)$ is a core for Q.

Part 3 Now let $f \in \mathrm{Dom}(Q)$ be bounded and of compact support. Given $0 < s < 1$ let $f_s := f * k_s$, where $k_s \in C_c^\infty$ is a standard mollifier. The proof that $f_s \in C_c^\infty$ depends upon differentiating under the integral sign. We have

$$\|f - f_s\|_2^2 = \int_{\mathbf{R}^N} |\mathcal{F} f(y)|^2 |1 - (2\pi)^{N/2} \hat{k}(sy)|^2 \mathrm{d}^N y$$

which converges to zero as $s \to 0$ by the dominated convergence theorem. A similar argument establishes that

$$\lim_{s\to 0} Q_0(f - f_s) = \lim_{s\to 0} \int_{\mathbf{R}^N} y^2 |\mathcal{F} f(y)|^2 |1 - (2\pi)^{N/2} \hat{k}(sy)|^2 \mathrm{d}^N y = 0.$$

If $g_s := |f - f_s|^2$ then $\lim_{s\to 0} \|g_s\|_1 = 0$ and there exists a compact set K which contains the support of every g_s. Since $\|f_s\|_\infty \le \|f\|_\infty$ for all $s > 0$, it follows that $\|g_s\|_\infty \le 4\|f\|_\infty^2$ for all $s > 0$. We claim that these facts are sufficient to deduce that $\lim_{s\to 0} \int_{\mathbf{R}^N} g_s V \mathrm{d}^N x = 0$. In other words

$$\lim_{s\to 0} Q_1(f - f_s) = 0$$

and C_c^∞ is indeed a core for $Q = Q_0 + Q_1$.

Part 4 We now establish the above claim. Given $\varepsilon > 0$ there exists a decomposition $V\big|_K = V_1 + V_\infty$ such that $\|V_1\|_1 < \varepsilon$ and $V_\infty \in L^\infty$. We then have

$$\begin{aligned} 0 \le \int_{\mathbf{R}^N} g_s V \, d^N x &\le \|g_s\|_\infty \|V_1\|_1 + \|g_s\|_1 \|V_\infty\|_\infty \\ &\le 4\varepsilon \|f\|_\infty^2 + \|g_s\|_1 \|V_\infty\|_\infty. \end{aligned}$$

Therefore

$$\limsup_{s\to 0} \int_{\mathbf{R}^N} g_s V \, d^N x \le 4\varepsilon \|f\|_\infty^2$$

for all $\varepsilon > 0$, and the claim is proved. □

It is clear that the above theorem can be extended to potentials which are bounded below and in L^1_{loc} by simply adding a large enough positive constant. If, however, we are interested in potentials with negative local singularities, then much stronger local conditions on the potential are necessary in order to be able to prove that the form Q is closed. Our theorem below is enough to handle many typical examples, but is far from the most general case which can now be treated.

We say that a potential V lies in $L^p + L^\infty$ if one can write $V = V_p + V_\infty$, where $V_p \in L^p$ and $V_\infty \in L^\infty$. This decomposition is not unique, and, if it is possible at all, then one can arrange for $\|V_p\|_p$ to be as small as one chooses. There are two theorems of importance, both of which have modifications for other values of the dimension N.

Theorem 8.2.2 *If H is defined on $L^2(\mathbf{R}^3)$ by $Hf := -\overline{\Delta} f + Vf$, where $V \in L^2 + L^\infty$, then H is self-adjoint and bounded below with the same domain as $H_0 := -\overline{\Delta}$.*

Proof We first write $V = V_2 + V_\infty$, where $V_2 \in L^2$ and $V_\infty \in L^\infty$. Theorem 3.6.5 states that $\lim_{s\to\infty} \|V_2(H_0 + s)^{-1}\| = 0$. We also have

$$\|V_\infty(H_0 + s)^{-1}\| \le \|V_\infty\| \, \|(H_0 + s)^{-1}\| = s^{-1}\|V_\infty\|$$

for all $s > 0$, so

$$\lim_{s\to\infty} \|V(H_0 + s)^{-1}\| = 0.$$

For any $\alpha > 0$, however small, we conclude that if s is large enough then

$$\|Vf\| \le \alpha \|H_0 f\| + \alpha s \|f\|$$

for all $f \in \mathrm{Dom}(H)$. The proof is now completed by an application of Theorem 1.4.2. □

Although the above theorem is of great importance, the range of potentials to which it applies is smaller than that for the following theorem.

Theorem 8.2.3 *Let $N \geq 3$ and let Q be defined on $W^{1,2}(\mathbf{R}^N)$ by*

$$Q(f) := \int_{\mathbf{R}^N} \{|\nabla f|^2 + V|f|^2\} \mathrm{d}^N x,$$

where $V \in L^p + L^\infty$ for some $p > N/2$. Then Q is a closed semibounded form and hence is associated to a self-adjoint operator H which is bounded below.

Note It is customary to write $H := -\overline{\Delta} + V$, but it must be remembered that this is a quadratic form sum and not an operator sum. The domains of the operators $-\overline{\Delta}$ and V are both dense linear subspaces of L^2, but the intersection of their domains can be $\{0\}$. See Exercise 8.1.

Proof Given $\varepsilon > 0$ we may write $V = V_p + V_\infty$ where $\|V_p\|_p < \varepsilon$ and $V_\infty \in L^\infty$. We then put $A := |V_p|^{1/2}$ and $B := |V_p|^{1/2}\mathrm{sign}(V_p)$. It follows from the case $\alpha = \frac{1}{2}$ of Theorem 3.6.5 that if $\varepsilon > 0$ is sufficiently small, then

$$\begin{aligned} \int_{\mathbf{R}^N} |V||f|^2 \mathrm{d}^N x &\leq \|Af\|_2 \|Bf\|_2 + \|V_\infty\|_\infty \|f\|_2^2 \\ &\leq \frac{1}{2} Q_0(f) + c_0 \|f\|_2^2 \end{aligned}$$

for some $c_0 > 0$ and all $f \in W^{1,2}(\mathbf{R}^N)$. Therefore there exists a positive constant c such that

$$\frac{1}{2}\{Q_0(f) + c\|f\|_2^2\} \leq Q(f) + c\|f\|_2^2 \leq \frac{3}{2}\{Q_0(f) + c\|f\|_2^2\}$$

for all $f \in W^{1,2}(\mathbf{R}^N)$. Corollary 4.4.3 now implies that the form $f \to Q(f) + c\|f\|_2^2$ is closed and is associated with a non-negative self-adjoint operator, which is clearly equal to $H + c1$. □

Example 8.2.4 Let $N \geq 3$ and consider the potential $V(x) := -|x|^{-\alpha}$. If $N = 3$ then the condition of Theorem 8.2.2 only holds if $0 \leq \alpha < \frac{3}{2}$. However, the condition of Theorem 8.2.3 holds for all $N \geq 3$ and all $0 \leq \alpha < 2$. Hence $H := -\overline{\Delta} + V$ may be interpreted as a semibounded self-adjoint operator for such α. However, if $\alpha > 2$, then not only does Theorem 8.2.3 become inapplicable but any analysis of the Schrödinger operator must be entirely different. The point is that the quadratic form Q, defined at least for functions in C_c^∞ whose support does not include the origin, is unbounded below. Therefore all of the standard quadratic form theorems are inapplicable. Physically also the system changes when $\alpha > 2$: in such cases some quantum particles approaching from infinity are sucked into a 'hole' at the origin, moving faster and faster as they get closer to the origin.

Although this is not the place to discuss the physics of such systems, we can at least verify that the quadratic form is not bounded below. Let ϕ be a function in C_c^∞ of unit L^2 norm and with support disjoint from the origin. Let $s > 0$ and put $\phi_s(x) := s^{N/2}\phi(sx)$ for all $x \in \mathbf{R}^N$. Then $\|\phi_s\|_2 = 1$ and a straightforward calculation shows that

$$\begin{aligned} Q(\phi_s) &= \int_{\mathbf{R}^N} \{|\nabla\phi_s|^2 - |x|^{-\alpha}|\phi_s|^2\}\mathrm{d}^N x \\ &= c_1 s^2 - c_2 s^\alpha, \end{aligned}$$

where $c_i > 0$ for $i = 1, 2$. It follows that $\lim_{s\to\infty} Q(\phi_s) = -\infty$.

Although quadratic form techniques are not appropriate for the analysis of operators of the type just mentioned, it is possible to develop their spectral theory. Specifically Lemma 1.2.8 ensures the existence of a self-adjoint extension of the operator initially defined on $C_c^\infty(\mathbf{R}^N\backslash\{0\})$. The identification of the different self-adjoint extensions depends upon a study of appropriate boundary conditions at $x = 0$. □

Example 8.2.5 The following example describes the interaction of two quantum-mechanical particles each moving in $\mathbf{R}^3$, the instantaneous configuration of the pair being described by a state in $L^2(\mathbf{R}^6)$. We label points in $\mathbf{R}^6$ by pairs (x, y) where $x, y \in \mathbf{R}^3$ represent the positions of the two particles. Let $W : \mathbf{R}^3 \to \mathbf{R}$ be a bounded potential controlling the interactions between the particles. Assuming that the particles both have mass m, the behaviour of the system is governed by the Schrödinger operator

$$Hf := -\frac{1}{2m}\overline{\Delta}_x f - \frac{1}{2m}\overline{\Delta}_y f + W(x - y)f.$$

The operator H is essentially self-adjoint on Schwartz space $\mathcal{S}$ by a combination of Theorem 1.4.2 and Theorem 3.5.3. The potential

$V(x,y) := W(x-y)$ of H does not vanish as $|x|^2+|y|^2 \to \infty$, and the spectral analysis is best done in two stages.

If we change to the variables $u := (x+y)/2$ and $v := x-y$ then the operator takes the form

$$\begin{aligned} H &:= -\frac{1}{4m}\overline{\Delta}_u - \frac{1}{m}\overline{\Delta}_v + W(v) \\ &= H_u + H_v, \end{aligned}$$

where $H_u = -(1/4m)\overline{\Delta}_u$ describes the free motion of the centre of mass, which is of mass $2m$, while H_v describes the interactions of the two particles, and is associated with a particle of reduced or effective mass $m/2$. The following theorem indicates that the spectral analysis of H itself is not interesting. □

Theorem 8.2.6 *There exists a constant $c \in \mathbf{R}$ such that* $\mathrm{Spec}(H) = [c,\infty)$.

Proof Let c be the smallest number in the spectrum of H_v. Let $f \in C_c^\infty(\mathbf{R}^6)$ have unit L^2 norm. Then

$$\begin{aligned} Q(f) &= \int_{\mathbf{R}^6} \left\{ \frac{1}{4m}|\nabla_u f|^2 + \frac{1}{m}|\nabla_v f|^2 + W(v)|f|^2 \right\} \mathrm{d}^3u\,\mathrm{d}^3v \\ &\geq \int_{\mathbf{R}^6} \left\{ \frac{1}{m}|\nabla_v f|^2 + W(v)|f|^2 \right\} \mathrm{d}^3u\,\mathrm{d}^3v \\ &\geq \int_{\mathbf{R}^3} c\|f(u,\cdot)\|_2^2\,\mathrm{d}^3u \\ &= c\|f\|_2^2. \end{aligned}$$

It follows that $\mathrm{Spec}(H) \subseteq [c,\infty)$.

The proof of the converse inclusion depends upon the use of Lemma 4.1.2. Let $\varepsilon > 0$ and let $h \in \mathrm{Dom}(H_v) \subseteq W^{1,2}(\mathbf{R}^3)$ have norm 1 and satisfy $\|H_v h - ch\|_2 \leq \varepsilon$. Let $d \geq 0$ and let $g \in \mathrm{Dom}(H_u) \subseteq W^{1,2}(\mathbf{R}^3)$ have norm 1 and satisfy $\|H_u g - dg\|_2 \leq \varepsilon$; such a g exists since $\mathrm{Spec}(H_u) = [0,\infty)$. The function $f(u,v) := g(u)h(v)$ lies in $\mathrm{Dom}(H)$ and

$$Hf(u,v) - (c+d)f(u,v) = \{H_u g(u) - dg(u)\}\, h(v) + g(u)\, \{H_v h(v) - ch(v)\}$$

for all $u, v \in \mathbf{R}^3$. Therefore

$$\|Hf - (c+d)f\|_2^2 \leq 2\varepsilon$$

and we deduce that $c+d \in \mathrm{Spec}(H)$. Since $d \geq 0$ is arbitrary, it follows that $[c,\infty) \subseteq \mathrm{Spec}(H)$. □

The operator H which we considered above is actually only appropriate for distinguishable spin zero particles of equal mass, but analogous calculations can be carried out for any number of particles of varying masses. After removing the centre of mass motion any negative eigenvalues of the remaining Schrödinger operator (H_v in our above example) describe particular bound state energies of a molecule composed of all of the particles.

The hydrogen atom, for an electron of mass $\frac{1}{2}$ and charge $-e$ moving around a fixed nucleus of charge $+e$, is described by the Schrödinger operator

$$Hf(x) := -\Delta f(x) - \frac{e^2}{|x|} f(x)$$

acting on $L^2(\mathbf{R}^3)$. It can be shown that the eigenvalues of H are all of the form $\lambda = -e^4/4n^2$, where n is a positive integer. The derivation of this formula, which was already known experimentally, was one of the first triumphs of quantum theory when it was formulated in 1925–6.

The corresponding result for the helium atom is not exactly soluble. The Schrödinger operator

$$H := -\Delta_{x_1} - \Delta_{x_2} - 2e^2 \left(\frac{1}{|x_1|} + \frac{1}{|x_2|} \right) + \frac{e^2}{|x_1 - x_2|} \tag{8.2.1}$$

acts on $L^2(\mathbf{R}^6)$ with $x = (x_1, x_2) \in \mathbf{R}^6$, and describes the motion of two electrons of mass $\frac{1}{2}$ and charge $-e$ moving around a fixed nucleus of charge $+2e$. The ground state energy of H was determined numerically by Hylleraas in 1930, using the Rayleigh–Ritz variational method, long before the advent of computers.

Example 8.2.7 The following is a simple case of an n-body Schrödinger operator, that is a Schrödinger operator describing the interaction of n particles of equal mass. We label points in $\mathbf{R}^{3n}$ by $x = (x_1, ..., x_n)$, where $x_r \in \mathbf{R}^3$ describes the position of the rth particle. We then write

$$Hf(x) := -\sum_{r=1}^{n} \overline{\Delta}_{x_r} f(x) + \sum_{1 \le r < s \le n} V(x_r - x_s) f(x)$$

and assume that $V \in L^p(\mathbf{R}^3) + L^\infty(\mathbf{R}^3)$ for some $p > \frac{3}{2}$. The associated quadratic form is

$$Q(f) := \int_{\mathbf{R}^{3n}} \{ |\nabla f|^2 + W|f|^2 \} \mathrm{d}^{3n}x, \tag{8.2.2}$$

where

$$W(x) := \sum_{1\le r<s\le n} V(x_r - x_s).$$

Even if V is of compact support the potential W does not converge to zero as $|x| \to \infty$. □

Theorem 8.2.8 *The quadratic form defined on $W^{1,2}(\mathbf{R}^{3n})$ by (8.2.2) is closed and semibounded, and hence is associated with a self-adjoint operator H which is bounded below.*

Proof After making a suitable orthogonal rotation in $\mathbf{R}^{3n}$ dependent upon r, s, we can prove as in Theorem 8.2.3 that

$$\int_{\mathbf{R}^{3n}} |V(x_r - x_s)||f(x)|^2 \mathrm{d}^{3n}x \le \frac{1}{n(n-1)} Q_0(f) + c\|f\|_2^2.$$

Summing over all r, s such that $1 \le r < s \le n$ we obtain

$$\int_{\mathbf{R}^{3n}} |W||f|^2 \mathrm{d}^{3n}x \le \frac{1}{2} Q_0(f) + \frac{n(n-1)c}{2} \|f\|_2^2,$$

after which the proof again follows Theorem 8.2.3. □

8.3 The positive spectrum

We find criteria on its potential for a Schrödinger operator to have spectrum which contains or is contained in the interval $[0, \infty)$. These criteria are quite different, and there are many Schrödinger operators which have neither property. We assume that the potentials V satisfy one of the hypotheses of Theorem 8.2.1, Theorem 8.2.3 or Example 8.2.7.

Theorem 8.3.1 *Suppose that there is a sequence of balls $B_m := B(a_m, r_m)$ with centres a_m and radii r_m, such that $\lim_{m\to\infty} r_m = +\infty$ and $\lim_{m\to+\infty} \| V\big|_{B_m} \|_\infty = 0$. Then $[0, \infty) \subseteq \mathrm{Spec}(H)$.*

Proof Let $\psi \in C_c^\infty$ be a function with support in $B(0, 1)$ and put $\psi_m(x) := \psi((x - a_m)/r_m)$ and $f_m(x) := \mathrm{e}^{iw\cdot x}\psi_m(x)$ where $w \in \mathbf{R}^N$. Then $\mathrm{supp}(f_m) \subseteq B_m$ and

$$Hf_m = |w|^2 \mathrm{e}^{iw\cdot x}\psi_m - 2i\mathrm{e}^{iw\cdot x} w \cdot \nabla\psi_m - \mathrm{e}^{iw\cdot x}\Delta\psi_m + V\mathrm{e}^{iw\cdot x}\psi_m.$$

Therefore

$$\|Hf_m - |w|^2 f_m\|_2 \le 2\|w \cdot \nabla\psi_m\|_2 + \|\Delta\psi_m\|_2 + \|V|_{B_m}\|_\infty \|\psi_m\|_2$$

and

$$\lim_{m\to\infty} \|Hf_m - |w|^2 f_m\|_2 / \|f_m\|_2 \le \lim_{m\to\infty} \left(c_1 r_m^{-1} + c_2 r_m^{-2} + \|V|_{B_m}\|_\infty\right) = 0.$$

This implies that $|w|^2 \in \mathrm{Spec}(H)$ by Lemma 4.1.2. The proof is completed by letting w range over the whole of $\mathbf{R}^N$. □

Corollary 8.3.2 *If H is the n-body Schrödinger operator of Example 8.2.7 and* $\lim_{|x|\to\infty} V(x) = 0$, *then* $[0, \infty) \subseteq \mathrm{Spec}(H)$.

Proof We need to consider a configuration in which the distance between every pair of particles increases to ∞. Let u be a unit vector in $\mathbf{R}^3$ and put

$$a_m := (mu, 2mu, ..., nmu) \in \mathbf{R}^{3n}$$

and $r_m := m/3$. It may then be seen that if $x := (x_1, ..., x_n) \in B(a_m, r_m)$ we have $|x_r - x_s| \ge m/3$ for all $r \ne s$. Therefore the hypothesis of Theorem 8.3.1 is satisfied. □

In our next few results we express a potential as the difference of its positive and negative parts: $V = V_+ - V_-$.

Lemma 8.3.3 *If*

$$\int_{\mathbf{R}^N} V_- |f|^2 \mathrm{d}^N x \le \int_{\mathbf{R}^N} |\nabla f|^2 \mathrm{d}^N x$$

for all $f \in C_c^\infty$, *then* $\mathrm{Spec}(H) \subseteq [0, \infty)$.

Proof We observe that $Q(f) \ge 0$ for all $f \in C_c^\infty$ and that such f form a core for Q, for reasons which depend upon which assumption we make on V. The lemma now follows by an application of the variational formula. □

Theorem 8.3.4 *If there exists a function* $0 < f \in C^\infty$ *for which*

$$V(x) \ge \{\Delta f(x)\}/f(x)$$

almost everywhere in $\mathbf{R}^N$, *then* $\mathrm{Spec}(H) \subseteq [0, \infty)$.

Proof Put $W := (\Delta f)/f$ and let $\phi \in C_c^\infty$. If we put $\psi := \phi/f$, then

$$\begin{aligned} Q(\phi) &\geq \int_{\mathbf{R}^N} \{-\psi\Delta f - 2\nabla f \cdot \nabla\psi - f\Delta\psi + Wf\psi\} f\overline{\psi}\, \mathrm{d}^N x \\ &= \int_{\mathbf{R}^N} \{-\overline{\psi}\nabla(f^2) \cdot \nabla\psi - (\Delta\psi)(f^2\overline{\psi})\} \mathrm{d}^N x \\ &= \int_{\mathbf{R}^N} \{-\overline{\psi}\nabla(f^2) \cdot \nabla\psi + \nabla\psi \cdot \nabla(f^2\overline{\psi})\} \mathrm{d}^N x \\ &= \int_{\mathbf{R}^N} f^2 |\nabla\psi|^2 \mathrm{d}^N x \geq 0. \end{aligned}$$

Since C_c^∞ is a form core, the result follows. □

Corollary 8.3.5 *If* $N \geq 3$ *and* $V_-(x) \leq (N-2)^2/4|x|^2$ *almost everywhere, then* $\mathrm{Spec}(H) \subseteq [0, \infty)$.

Note This corollary is trivial for $N = 2$ and false for $N = 1$.

Proof Put $f(x) := (s^2 + |x|^2)^\alpha$, where $\alpha \in \mathbf{R}$ and $s > 0$. A direct computation shows that

$$(\Delta f(x))/f(x) = \frac{(4\alpha^2 + 2\alpha(N-2))|x|^2 + 2N\alpha s^2}{(s^2 + |x|^2)^2}$$

for all $x \in \mathbf{R}^N$. If we put $\alpha := -(N-2)/4$ and

$$W_s(x) := \frac{(N-2)^2|x|^2}{4(s^2 + |x|^2)^2} + \frac{N(N-2)s^2}{2(s^2 + |x|^2)^2},$$

then Theorem 8.3.4 shows that

$$\int_{\mathbf{R}^N} W_s |\phi|^2 \mathrm{d}^N x \leq \int_{\mathbf{R}^N} |\nabla\phi|^2 \mathrm{d}^N x$$

for all $\phi \in C_c^\infty$. If $N \geq 3$, then $W_s(x) > 0$ for all $x \in \mathbf{R}^N$. Letting $s \to 0$, Fatou's lemma now establishes that

$$\int_{\mathbf{R}^N} \frac{(N-2)^2}{4|x|^2} |\phi|^2 \mathrm{d}^N x \leq \int_{\mathbf{R}^N} |\nabla\phi|^2 \mathrm{d}^N x$$

for all $\phi \in C_c^\infty$, and the proof is completed by an application of Lemma 8.3.3. □

8.4 Compact perturbations

This section is devoted to some functional analytic results concerning the essential spectrum which will be needed in Section 8.5. We shall use the spectral subspaces $L_{(a,b)} := P_{(a,b)}\mathcal{H}$ defined in Section 2.5, and the characterization of the essential spectrum EssSpec(H) given in Lemma 4.1.4.

Lemma 8.4.1 *Let H be a self-adjoint operator acting on the Hilbert space $\mathcal{H}$ and let $\lambda \in \mathbf{R}$. The following are equivalent:*

(1) $\lambda \in \mathrm{EssSpec}(H)$.

(2) *For all $\varepsilon > 0$ and $n > 0$ there exists a subspace $L_{n,\varepsilon} \subseteq \mathrm{Dom}(H)$ such that $\dim(L_{n,\varepsilon}) \geq n$ and $\|Hf - \lambda f\| \leq \varepsilon\|f\|$ for all $f \in L_{n,\varepsilon}$.*

(3) *For all $\varepsilon > 0$ there exists a subspace $L_\varepsilon \subseteq \mathrm{Dom}(H)$ such that $\dim(L_\varepsilon) = \infty$ and $\|Hf - \lambda f\| \leq \varepsilon\|f\|$ for all $f \in L_\varepsilon$.*

Proof The implication (1) $\Rightarrow$ (3) follows from Lemma 4.1.4 upon putting $L_\varepsilon := L_{(\lambda-\varepsilon,\lambda+\varepsilon)}$. Since (3) $\Rightarrow$ (2) is trivial, we are left with proving that (2) $\Rightarrow$ (1).

Suppose that $L_{(\lambda-\varepsilon,\lambda+\varepsilon)}$ has finite dimension for some $\varepsilon > 0$ and let n be an integer larger than this dimension. Let P be the orthogonal projection from $\mathcal{H}$ onto $L_{(\lambda-\varepsilon,\lambda+\varepsilon)}$. Dimension counting shows that P cannot be one-one from $L_{n,\varepsilon/2}$ to $L_{(\lambda-\varepsilon,\lambda+\varepsilon)}$. Therefore there exists a non-zero element f of $L_{n,\varepsilon/2}$ which is orthogonal to $L_{(\lambda-\varepsilon,\lambda+\varepsilon)}$. An application of the spectral theorem shows that $\|Hf - \lambda f\| \geq \varepsilon\|f\|$. The contradiction establishes that the hypothesis at the start of this paragraph is false, and hence that $\lambda \in \mathrm{EssSpec}(H)$. □

Lemma 8.4.2 *If $H \geq 0$ and $\lambda \geq 0$ then $\lambda \in \mathrm{EssSpec}(H)$ if and only if $(\lambda+1)^{-1} \in \mathrm{EssSpec}((H+1)^{-1})$.*

Proof This is an immediate consequence of Lemma 4.1.4 and the spectral theorem. □

Theorem 8.4.3 *Let H and K be two non-negative self-adjoint operators and suppose that*

$$A := (H+1)^{-1} - (K+1)^{-1}$$

is compact. Then H and K have the same essential spectrum.

Proof If $\lambda \in \mathrm{EssSpec}(H)$ and $\varepsilon > 0$, then there exists an infinite-dimensional subspace L of $\mathcal{H}$ such that

$$\|(H+1)^{-1}f - (\lambda+1)^{-1}f\| \leq \varepsilon\|f\|$$

for all $f \in L$. Since $(H+1)^{-1}$ is bounded, there is no loss in assuming that L is closed. Let $B := PA^2P|_L$, where P is the orthogonal projection onto L. Then B is compact and self-adjoint and an application of Theorem 4.2.2 to B shows that there exists an infinite-dimensional subspace M of L such that $\|Bf\| \leq \varepsilon^2\|f\|$ for all $f \in M$. It follows that $\|Af\| \leq \varepsilon\|f\|$ for all $f \in M$. Therefore

$$\|(K+1)^{-1}f - (\lambda+1)^{-1}f\| \leq 2\varepsilon\|f\|$$

for all $f \in M$, and we deduce by Lemma 8.4.1 that $\lambda \in \text{EssSpec}(K)$. The reverse inclusion is proved by a similar argument. □

8.5 The negative spectrum

We discuss several different results giving information about the negative spectrum. The results presented here can be proved under much weaker assumptions, and our only goal is to indicate the kind of behaviour which can occur.

Theorem 8.5.1 *Let $N \geq 3$ and let $V \in L^p$ for some $p > N/2$. The operator $H := -\overline{\Delta}+V$ has essential spectrum equal to $[0,\infty)$. If H has any negative spectrum, then it consists of a finite sequence of negative eigenvalues of finite multiplicity, or an infinite sequence which converges to zero.*

Proof If we put $H_0 := -\overline{\Delta}$ then $\text{Dom}(H_0^{1/2}) = \text{Dom}(|H|^{1/2}) = W^{1,2}$ by Theorem 8.2.3. If $c > 0$ is large enough, then

$$\frac{1}{2}\|(H_0+c)^{1/2}f\|_2 \leq \|(H+c)^{1/2}f\|_2 \leq \frac{3}{2}\|(H_0+c)^{1/2}f\|_2$$

for all $f \in W^{1,2}$ and both norms are equivalent to the standard norm of $W^{1,2}$. Now

$$\begin{aligned}(H_0+c)^{-1} - (H+c)^{-1} &= (H_0+c)^{-1}V(H+c)^{-1}\\ &= A\,B\,C\,D\,E,\end{aligned}$$

where $A := (H_0+c)^{-1/2}$, $B := (H_0+c)^{-1/2}|V|^{1/2}$, $C := \text{sign(V)}|V|^{1/2}(H_0+c)^{-1/2}$, $D := (H_0+c)^{1/2}(H+c)^{-1/2}$ and $E := (H+c)^{-1/2}$. It is clear that A and E are bounded operators on L^2. Moreover, B^* and C are compact operators on L^2 by Corollary 3.6.6. In order to see that D is a bounded linear operator on L^2 we observe that $(H+c)^{-1/2}$ is a bounded operator from L^2 to $\text{Dom}(|H|^{1/2}) = W^{1,2}$ and $(H_0+c)^{1/2}$ is a bounded operator from $\text{Dom}(H_0^{1/2}) = W^{1,2}$ to L^2.

Since it is a product of bounded and compact operators, the difference of the two resolvents is compact. It now follows from Theorem 8.4.3 that H and H_0 have the same essential spectrum, and the proof of the first statement of the theorem is completed by observing that the essential spectrum of H_0 equals $[0,\infty)$. Any negative point in the spectrum must be an eigenvalue of finite multiplicity. Since H is semibounded and any limit of negative eigenvalues lies in the essential spectrum, the only possible limit is 0. □

Theorem 8.5.2 *Suppose that $N \geq 3$, that $V \in L^p$ for some $p > N/2$, that V is not identically zero, and that it satisfies a bound of the type*

$$0 \leq V(x) \leq \sum_{r=1}^{n} c_r |x - a_r|^{-2}$$

for all $x \in \mathbf{R}^N$, where $c_r \geq 0$ and $a_r \in \mathbf{R}^N$ for all r. Given $0 < \lambda < \infty$, let E_λ denote the bottom of the spectrum of $H_\lambda := -\overline{\Delta} - \lambda V$. Then E_λ is a monotonically decreasing concave function of λ and there exists a 'critical threshold' $\lambda_c > 0$ such that $E_\lambda = 0$ if $\lambda < \lambda_c$ but $E_\lambda < 0$ otherwise.

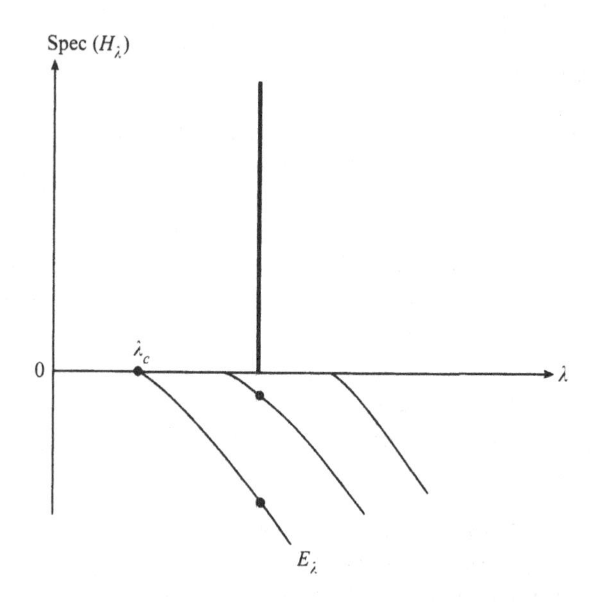

The heavy line in the diagram together with the two points is the spectrum of H_λ for a typical value of λ.

Proof We first observe that

$$E_\lambda = \inf\{Q_\lambda(f) : f \in C_c^\infty \text{ and } \|f\|_2 = 1\},$$

because C_c^∞ is a form core for Q_λ. Let $\lambda = (1-t)\alpha + t\beta$ where $\alpha > 0$, $\beta > 0$ and $0 < t < 1$. Given $\varepsilon > 0$ there exists $f \in C_c^\infty$ with $\|f\|_2 = 1$ such that $E_\lambda \le Q_\lambda(f) \le E_\lambda + \varepsilon$. Therefore

$$\begin{aligned}(1-t)E_\alpha + tE_\beta &\le (1-t)Q_\alpha(f) + tQ_\beta(f)\\ &= Q_\lambda(f) \le E_\lambda + \varepsilon.\end{aligned}$$

Letting $\varepsilon \to 0$ the concavity of the function $\lambda \to E_\lambda$ follows.

Since $H_\lambda \le H_0 = -\overline{\Delta}$ for all $\lambda > 0$ it follows that $E_\lambda \le 0$ for such λ. The fact that $E_\lambda = 0$ for small $\lambda > 0$ follows from Corollary 8.3.5. If $f \in C_c^\infty$ has $\|f\|_2 = 1$, then

$$\begin{aligned}E_\lambda &\le Q_\lambda(f)\\ &= \int_{\mathbf{R}^N} \{|\nabla f|^2 - \lambda V|f|^2\} \mathrm{d}^N x\\ &= k_1 - k_2\lambda,\end{aligned}$$

where $k_i > 0$. Therefore $\lim_{\lambda\to\infty} E_\lambda = -\infty$. Finally if $\alpha < \beta$, then $Q_\alpha(f) \ge Q_\beta(f)$ for all $f \in C_c^\infty$ and we deduce that E_λ is a monotone decreasing function of λ. □

Some aspects of the above theorem can be extended to potentials of the assumed type which have both positive and negative parts.

Theorem 8.5.3 *Let $N \ge 3$, $H_\lambda := -\overline{\Delta} + \lambda V$ where $0 < \lambda < \infty$ and $V \in L^p$ for some $p > N/2$. Suppose further that there exists an open set $U \subseteq \mathbf{R}^N$ on which V is negative. If E_λ is the bottom of the spectrum of H_λ, then $E_\lambda \le 0$ for all $\lambda \ge 0$ and $\lim_{\lambda\to\infty} E_\lambda = -\infty$.*

Proof We use Theorem 8.5.1 to prove that $E_\lambda \le 0$ for all $\lambda \ge 0$. The function f used in Theorem 8.5.2 to prove that $\lim_{\lambda\to\infty} E_\lambda = -\infty$ must be chosen to have support in U. □

The following results correctly suggest that the crucial issue for the existence of negative eigenvalues for all $\lambda > 0$, however small, is the rate at which the potential $V(x)$ converges to 0 as $|x| \to \infty$.

Example 8.5.4 Let $N \ge 3$ and let $H := -\Delta - \lambda V$ where $V(x) := (|x|^2+1)^{-\gamma}$ for some $\gamma > 0$ and $\lambda > 0$. If $\gamma \ge 1$ then Theorem 8.5.2 is applicable and there exists a positive threshold for the existence of negative eigenvalues. If $0 < \gamma < 1$ the situation is totally different. □

Theorem 8.5.5 *If $0 < \gamma < 1$ in Example 8.5.4, then H has non-empty negative spectrum for all $\lambda > 0$.*

Proof Let $f(x) := (s^2 + |x|^2)^{-N}$, where $s \geq 1$. Direct calculations show that $f \in W^{1,2}$ and that $W := -(\Delta f)/f$ is given by

$$W(x) = \frac{2N^2}{s^2 + |x|^2} - \frac{4N(N+1)|x|^2}{(s^2 + |x|^2)^2}.$$

If $\lambda > 0$ and $0 < \gamma < 1$, there exists $s > 1$ such that

$$W(x) \leq \frac{2N^2}{s^2 + |x|^2} < \frac{\lambda}{(1 + |x|^2)^\gamma} = \lambda V(x)$$

for all $x \in \mathbf{R}^N$. Hence

$$\int_{\mathbf{R}^N} \{|\nabla f|^2 - \lambda V|f|^2\} \mathrm{d}^N x < \int_{\mathbf{R}^N} \{|\nabla f|^2 - W|f|^2\} \mathrm{d}^N x = 0$$

and an application of the Rayleigh–Ritz formula yields $E_\lambda \leq Q_\lambda(f)/\|f\|_2^2 < 0$. □

8.6 Double wells

In this final section we consider double well Schrödinger operators acting on $L^2 := L^2(\mathbf{R}^3)$. One such operator is

$$H_r f(x) := -\overline{\Delta} f(x) + A(x - re)f(x) + B(x + re)f(x),$$

where A, B are two potentials on $\mathbf{R}^3$ and $e := (1, 0, 0)$. We study the spectral behaviour as $r \to \infty$, assuming that $A, B \in L^2(\mathbf{R}^3) + L^\infty(R^3)$, $\lim_{|x|\to\infty} A(x) = 0$ and $\lim_{|x|\to\infty} B(x) = 0$. The analysis can be extended to higher dimensions and more general potentials with little difficulty.

The operator H_r is compared with $H_0 := -\overline{\Delta}$, $H_A := -\overline{\Delta} + A$ and $H_B := -\overline{\Delta} + B$. All of these operators have domain $W^{2,2}(\mathbf{R}^3)$ by Theorem 8.2.2 and essential spectrum $[0, \infty)$ by Theorem 8.5.1. If $N(r)$, $N(A)$ and $N(B)$ denote the negative spectrum of H_r, H_A and H_B respectively (including 0 in each case), then we show that $N(r)$ converges as a set to $N(A) \cup N(B)$ as $r \to \infty$.

It can be shown under quite weak conditions that the convergence is very rapid, and precise information about the asymptotic behaviour can be obtained. Our method can be extended to provide some information of this type, but its main merit lies in its technical simplicity.

We consider two operators, namely

$$K_r := \begin{pmatrix} H_r & 0 \\ 0 & H_0 \end{pmatrix} \quad , \quad K := \begin{pmatrix} H_A & 0 \\ 0 & H_B \end{pmatrix},$$

both acting on $\mathscr{H} := L^2 \oplus L^2$. The negative spectrum of K_r equals that of H_r, while the negative spectrum of K equals $N(A) \cup N(B)$. Our claim concerning the convergence of the negative spectrum follows from Corollary 2.6.3 as soon as we establish the following result.

Theorem 8.6.1 *There exists a family of unitary operators U_r on $\mathscr{H}$ such that $U_r^* K_r U_r$ converge in the norm resolvent sense to K as $r \to \infty$.*

Proof This involves making a number of constructions. Define the unitary operator $V_r : \mathscr{H} \to \mathscr{H}$ by

$$V_r \begin{pmatrix} f(x) \\ g(x) \end{pmatrix} = \begin{pmatrix} \cos\{\sigma(x_1/r)\} & \sin\{\sigma(x_1/r)\} \\ -\sin\{\sigma(x_1/r)\} & \cos\{\sigma(x_1/r)\} \end{pmatrix} \begin{pmatrix} f(x) \\ g(x) \end{pmatrix}$$

where $\sigma : \mathbf{R} \to [0, \pi/2]$ is defined by

$$\sigma(s) = \begin{cases} \dfrac{\pi}{2} & \text{if} \quad s \le -\dfrac{1}{3} \\ \dfrac{\pi}{4}(1-3s) & \text{if} \quad -\dfrac{1}{3} < s < \dfrac{1}{3} \\ 0 & \text{if} \quad s \ge \dfrac{1}{3}. \end{cases},$$

It may be seen that $V_r(x) = \begin{pmatrix} 1 & 0 \\ 0 & 1 \end{pmatrix}$ if $x_1 \ge r/3$, while $V_r(x) = \begin{pmatrix} 0 & 1 \\ -1 & 0 \end{pmatrix}$ if $x_1 \le -r/3$. Our calculations will be aided by defining the following functions, also regarded as multiplication operators on L^2, namely

$$\begin{aligned} C_r(x) &:= \cos\{\sigma(x_1/r)\}, \\ S_r(x) &:= \sin\{\sigma(x_1/r)\}, \\ A_r(x) &:= A(x - re), \\ B_r(x) &:= B(x + re). \end{aligned}$$

We also let D_i denote the operator $\partial/\partial x_i$. The central formula upon which the proof of the theorem is based is

$$V_r K_r V_r^* = \begin{pmatrix} H_0 + A_r & 0 \\ 0 & H_0 + B_r \end{pmatrix} + D_1 F_r - F_r D_1 + G_r. \tag{8.6.1}$$

In this formula F_r and G_r are bounded matrix-valued functions on $\mathbf{R}^3$.

The proof of (8.6.1) involves standard matrix manipulations together with the use of a number of elementary identities. These include $\Delta = D_1^2 + D_2^2 + D_3^2$, $[D_1, S_r] = S_r'$, $[D_2, S_r] = 0$ and $[D_3, S_r] = 0$. An idea of the effect of this unitary operator is given by the following diagram.

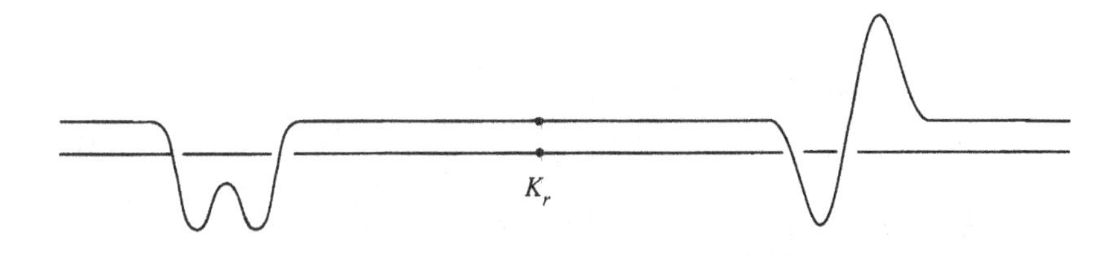

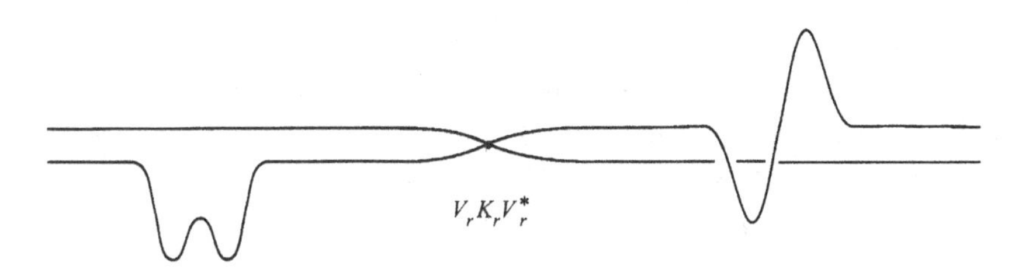

The next step is to observe that

$$\lim_{r\to\infty} \|F_r\| = 0 \quad , \quad \lim_{r\to\infty} \|G_r\| = 0.$$

The proof of these formulae involves observing that $S_r A_r$, $C_r B_r$, S_r' and C_r' all converge in norm to 0 as $r \to \infty$.

The final construction involves using space translations to change the potentials A_r and B_r into A and B. We define the unitary operators W_r on $\mathscr{H}$ by

$$W_r := \begin{pmatrix} T_r & 0 \\ 0 & T_r^* \end{pmatrix},$$

where $T_r f(x) := f(x + re)$ for all $f \in L^2$. Applying this to (8.6.1) we obtain

$$W_r V_r K_r V_r^* W_r^*$$
$$= \begin{pmatrix} H_0 + A & 0 \\ 0 & H_0 + B \end{pmatrix} + D_1 W_r F_r W_r^* - W_r F_r W_r^* D_1 + W_r G_r W_r^*.$$

This may be rewritten in the form

$$U_r K_r U_r^* = K + D_1 \tilde{F}_r - \tilde{F}_r D_1 + \tilde{G}_r, \qquad (8.6.2)$$

where U_r are unitary, while $\tilde{F}_r$ and $\tilde{G}_r$ converge in norm to 0 as $r \to \infty$.

We can now finally prove the norm resolvent convergence claimed in the theorem. From (8.6.2) we deduce that

$$(K+i)^{-1} - (U_r K_r U_r^* + i)^{-1} = (K+i)^{-1}(D_1 \tilde{F}_r - \tilde{F}_r D_1 + \tilde{G}_r)(U_r K_r U_r^* + i)^{-1},$$

so

$$\begin{aligned} &\|(K+i)^{-1} - (U_r K_r U_r^* + i)^{-1}\| \\ &\quad \leq \|(K+i)^{-1} D_1\| \, \|\tilde{F}_r\| + \|\tilde{F}_r\| \, \|D_1 (U_r K_r U_r^* + i)^{-1}\| + \|G_r\|. \qquad (8.6.3) \end{aligned}$$

Since K has domain $W^{2,2}$, $(K+i)^{-1}$ is bounded from $\mathscr{H}$ to $W^{2,2}$. Since D_1 is bounded from $W^{2,2}$ to $\mathscr{H}$ it follows that

$$\begin{aligned} \|(K+i)^{-1} D_1\| &= \|D_1 (K-i)^{-1}\| \\ &< \infty. \end{aligned}$$

Now $(K_r + i)^{-1}$ are uniformly bounded from $\mathscr{H}$ to $W^{2,2}$ for $r \geq 1$, W_r and V_r are uniformly bounded from $W^{2,2}$ to $W^{2,2}$ for $r \geq 1$, and D_1 is bounded from $W^{2,2}$ to $\mathscr{H}$. Therefore

$$\begin{aligned} \|D_1 (U_r K_r U_r^* + i)^{-1}\| &= \|D_1 U_r (K_r + i)^{-1} U_r^*\| \\ &= \|D_1 W_r V_r (K_r + i)^{-1}\| \\ &\leq c \end{aligned}$$

for some $c < \infty$ and all $r \geq 1$. Applying these bounds to (8.6.3) we finally deduce that

$$\lim_{r \to \infty} \|(K+i)^{-1} - (U_r K_r U_r^* + i)^{-1}\| = 0. \qquad \square$$

It is not easy to find simple examples to illustrate the above theorem. The simplest case is that of a double delta function potential in one dimension. This does not satisfy the technical conditions of our theorem above, but it can be solved exactly, so no theorems are actually necessary.

Example 8.6.2 We consider the operator on $L^2(\mathbf{R})$ given formally by

$$H_r := -\frac{\mathrm{d}^2}{\mathrm{d}x^2} - 2\delta_r - 2\delta_{-r}. \qquad (8.6.4)$$

More precisely the operator is given by $Hf(x) := -f''(x)$ for all f in its domain $\mathscr{D}_r$, which we define as follows. We require that every function f in $\mathscr{D}_r$ should be twice continuously differentiable on $\mathbf{R}$ except that f'

and f'' are allowed to have simple jump discontinuities at $\pm r$. At these points the boundary conditions

$$f'(r+) - f'(r-) = -2f(r),$$
$$f'(-r+) - f'(-r-) = -2f(-r)$$

must be satisfied. Finally f, f' and f'' should all be $O(|x|^{-1})$ as $|x| \to \infty$. The verification that H_r is symmetric on this domain is straightforward. The fact that it is also bounded below uses Theorem 3.7.6. It is actually also essentially self-adjoint, but we ignore this and define the associated self-adjoint operator to be the Friedrichs extension.

The reason for the rather strange formula (8.6.4) for the operator H_r is that its quadratic form is given by

$$Q_r(f) := \int_{-\infty}^{\infty} |f'(x)|^2 \mathrm{d}x - 2|f(-r)|^2 - 2|f(r)|^2$$

for all f in its domain, which is $W^{1,2}(\mathbf{R})$. See the note after Lemma 7.1.1 for the interpretation of this formula. A rigorous treatment of the quadratic form can be based upon Corollary 4.4.3.

The operator H_r has exactly two negative eigenvalues with eigenfunctions lying in D_r. The more negative eigenvalue corresponds to an even eigenfunction of the form

$$f(x) := \begin{cases} c\mathrm{e}^{\beta x} & \text{if} \quad x < -r, \\ \mathrm{e}^{\beta x} + \mathrm{e}^{-\beta x} & \text{if} \quad |x| \le r, \\ c\mathrm{e}^{-\beta x} & \text{if} \quad x > r. \end{cases}$$

The condition $f \in L^2$ forces $\beta > 0$, while the boundary conditions force

$$\mathrm{e}^{\beta r} + \mathrm{e}^{-\beta r} = c\mathrm{e}^{-\beta r}$$
$$c\beta\mathrm{e}^{-\beta r} + \beta\mathrm{e}^{\beta r} - \beta\mathrm{e}^{-\beta r} = 2c\mathrm{e}^{-\beta r}.$$

The solution of these equations is

$$\beta\,\{1 + \tanh(\beta r)\} = 2.$$

One sees that β decreases from 2 to 1 as r increases from 0 to ∞. The eigenvalue of H_r is given by $\lambda = -\beta^2$.

The less negative eigenvalue corresponds to an odd eigenfunction of the form

$$f(x) := \begin{cases} -c\mathrm{e}^{\gamma x} & \text{if} \quad x < -r, \\ \mathrm{e}^{\gamma x} - \mathrm{e}^{-\gamma x} & \text{if} \quad |x| \le r, \\ c\mathrm{e}^{-\gamma x} & \text{if} \quad x > r. \end{cases}$$

The condition $f \in L^2$ forces $\gamma > 0$, while the boundary conditions force

$$e^{\gamma r} - e^{-\gamma r} = ce^{-\gamma r},$$
$$c\gamma e^{-\gamma r} + \gamma e^{\gamma r} + \gamma e^{-\gamma r} = 2ce^{-\gamma r}.$$

The solution of these equations is

$$\gamma \left\{1 + \tanh(\gamma r)^{-1}\right\} = 2.$$

One sees that γ increases from 0 to 1 as r increases from $\frac{1}{2}$ to ∞. The eigenvalue of H_r is given by $\mu = -\gamma^2$.

The graphs of the two eigenfunctions are drawn below.

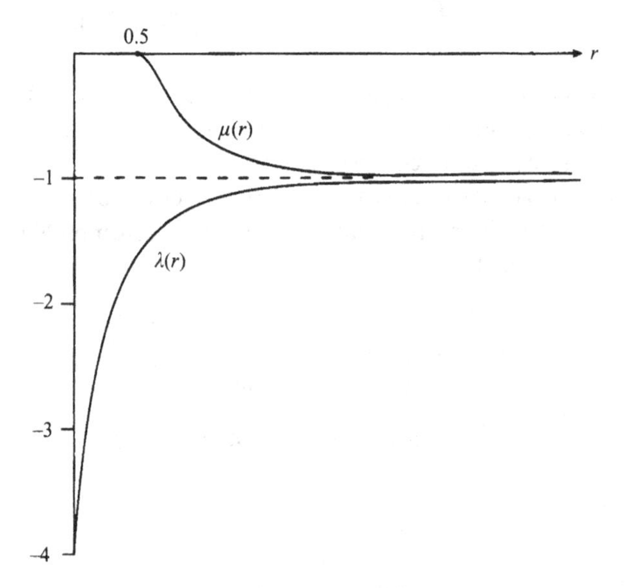

The first observation is that there is a threshold $r = \frac{1}{2}$ for the appearance of a second negative eigenvalue. One also sees that the two eigenvalues converge to -1 as $r \to \infty$ in conformity with Theorem 8.6.1. The convergence is rapid, and direct calculations show that

$$\mu(r) - \lambda(r) = 2e^{-2r}\left\{1 + o(1)\right\}$$

as $r \to \infty$. □

Exercises

8.1 Use Fourier transforms to prove that the domain of $H_0 := -\overline{\Delta}$ on $L^2(\mathbf{R}^3)$ consists entirely of continuous bounded functions. If V is a non-negative potential which is not L^2 when restricted to any non-empty bounded open subset of $\mathbf{R}^3$, prove that $\mathrm{Dom}(H_0) \cap \mathrm{Dom}(V) = \{0\}$.

8.2 Let $H := -\overline{\Delta} + V$ on $L^2(\mathbf{R}^N)$ and put

$$c := \liminf\{V(x) : |x| \to \infty\}.$$

Prove that the essential spectrum of H is contained in $[c, \infty)$.

8.3 Prove that if $H := -\overline{\Delta} + V$ on $L^2(\mathbf{R}^N)$ and $\lim_{|x|\to\infty} V(x) = +\infty$, then H has compact resolvent.

8.4 Find the Schrödinger operator on $L^2(\mathbf{R}^N)$ whose ground state is

$$\phi(x) := c_1 \exp[-c_2 \langle x \rangle^\alpha]$$

for given $\alpha > 0$, with particular attention to the cases $\alpha = 1, 2, 3$. What is the essential spectrum of such operators?

8.5 Find the Schrödinger operator on $L^2(\mathbf{R}^N)$ whose ground state is

$$\phi(x) := c\langle x \rangle^{-\alpha}$$

for given $\alpha > N/2$. What is the essential spectrum of such operators?

8.6 Let $H_{\lambda,\alpha}$ be defined on $L^2(\mathbf{R}^N)$ by

$$H_{\lambda,\alpha} := -\overline{\Delta} + \lambda|x|^\alpha$$

for some $\alpha > 0$ and $\lambda > 0$. Use a scaling argument to determine explicitly how the eigenvalues of $H_{\lambda,\alpha}$ depend upon λ.

8.7 Prove that the helium atom operator H of (8.2.1) is self-adjoint on the same domain $W^{2,2}(\mathbf{R}^6)$ in $L^2(\mathbf{R}^6)$ as the Laplacian by applying Theorem 8.2.2 separately to each of the terms in the potential.

8.8 Carry out similar calculations to those of Example 8.6.2 for the asymmetric Schrödinger operator

$$H := -\frac{\mathrm{d}^2}{\mathrm{d}x^2} - \delta_{-r} - 2\delta_r.$$

Bibliography

Adams R. A. (1975): *Sobolev Spaces*. Academic Press, New York.

Cycon H. L., Froese R. G., Kirsch W. and Simon B. (1987): *Schrödinger Operators, with Applications to Quantum Mechanics and Global Geometry*. Springer-Verlag, Berlin, Heidelberg and New York.

Davies E. B. (1980): *One-Parameter Semigroups*. Academic Press, New York.

Davies E. B. (1989): *Heat Kernels and Spectral Theory*. Cambridge University Press, Cambridge.

Davies E. B. (1995): The functional calculus, *J. London Math. Soc.* (2) **52**, 166–176.

Dieudonné J. (1981): *History of Functional Analysis*. North-Holland Mathematical Studies 49. North-Holland Publ. Co., Amsterdam.

Helffer B. and Sjöstrand J. (1989): Equation de Schrödinger avec Champ Magnetique et Equation de Harper. pp118–197 in *Schrödinger Operators*, eds. H. Holden and A. Jensen, Lecture Notes in Physics, Vol. 345. Springer-Verlag, Berlin, Heidelberg and New York.

Kato T. (1966): *Perturbation Theory of Linear Operators*. Springer-Verlag, Berlin, Heidelberg and New York.

Körner T. W. (1988): *Fourier Analysis*. Cambridge University Press, Cambridge.

Opic B. and Kufner A. (1990): *Hardy-type Inequalities*. Pitman Research Notes in Mathematics 219. Longman Scientific and Technical, Harlow.

Reed M. and Simon B. (1975): *Methods of Modern Mathematical Physics*. Vol2, Fourier Analysis. Academic Press, New York.

Reed M. and Simon B. (1978): *Methods of Modern Mathematical Physics*. Vol4, Analysis of Operators. Academic Press, New York.

Notation index

Index

Printed in the United States
By Bookmasters